Personality and Neurosurgery

Proceedings of the Third Convention of the Academia Eurasiana Neurochirurgica Brussels, August 30 – September 2, 1987

Edited by
J. Brihaye, L. Calliauw
F. Loew, R. van den Bergh

Acta Neurochirurgica
Supplementum 44

Springer-Verlag Wien New York

Prof. Dr. Jean Brihaye
Clinique Neurochirurgicale, Bruxelles, Belgium

Prof. Dr. Luc Calliauw
Kliniek voor Neurochirurgie, Gent, Belgium

Prof. Dr. Friedrich Loew
Neurochirurgische Universitätsklinik, Homburg/Saar, Federal Republic of Germany

Prof. Dr. Raymond van den Bergh
Kliniek voor Neurologie en Neurochirurgie, Katholieke Universiteit Leuven, Belgium

With 35 partly coloured Figures

Product Liability: The publisher can give no guarantee for information about drug dosage and application thereof contained in this book. In every individual case the respective user must check its accuracy by consulting other pharmaceutical literature.

Softcover reprint of the hardcover 1st edition 1988

ISSN 0065-1419

ISBN-13:978-3-7091-9007-4 e-ISBN-13:978-3-7091-9005-0
DOI: 10.1007/978-3-7091-9005-0

Preface

The human personality is inextricably bound up with, among other things, the function of the central nervous system. Diseases and malfunctions of the brain, head injuries and neurosurgical operations can all result in permanently altered behaviour patterns. This interrelation between brain and behaviour is most clearly demonstrated in cases involving functional neurosurgery and severe traumatic lesions.

Despite the fact that this interrelation represents an everyday challenge to the neurosurgeon, it is a question which receives less attention than it deserves in neurosurgical meetings.

Given the scope and complexity of this topic, it is not possible to cover every aspect of it here: hence, discussion is limited to the impact on personality of injuries, language, epilepsy and psychosurgery.

However, before considering the medical aspects, it was deemed necessary to try and arrive at a definition of "personality". This question was discussed by a number of philosophers representing various perspectives. Their diversity of viewpoints and conceptions greatly enriched the discussions.

Jean Brihaye   Luc Calliauw   Friedrich Loew   Raymond van den Bergh

Contents

Listed in Current Contents

VI. Aphasia and Personality

VII. Psychosurgery and Personality

VIII. General Conclusions

I. Introduction

Acta Neurochirurgica, Suppl. 44, 3–4 (1988)

Address

by His Excellency, Mr. **Yoshiya Kato**

Ambassador of Japan to the Kingdom of Belgium

Mr. President, Distinguished Guests,

It is indeed a great honour and a special privilege this evening for me to be given the opportunity of addressing such an eminent gathering of "wise men".

Although I am the merest layman in matters of medical science, you have seen fit to invite me here, for this honourable Academy of Neurological Surgery has always brought together, not only scientists and physicians, but also priests, theologians, philosophers and historians of different religions and nationalities. Why not a diplomat then? Indeed, problems of health have now become topics of diplomacy, in the same way as international politics and world economy, disarmament and foreign aid, terrorism and environment and so on. For example, most recently at the Industrial Summit in Venice, a Chairman's statement was issued on AIDS, urging international cooperation to cope with it, while at previous Summit meetings, biotechnology, cancers and drugs had been discussed among heads of state and government of major industrial nations.

I am particularly happy to note that this year's Convention takes place in this great and noble city of Brussels, following that held in Hakone, Japan, last year.

Belgium and Japan, it should be recalled, have long been linked by close historical ties in the field of medicine and surgery. The first Japanese text with illustrations on anatomy was published in Edo—present day Tokyo—in 1774, under the titel of "Kaitai Shinsho", or New Treatise of Anatomy. It was a translation of "Ontleedkundige Tabelen", the Dutch version of a German work by Johan Adam Kulmus. Kulmus himself had borrowed most of his illustrations from "Corporis humani Anatomia" by Philippe Verheyen, then professor at the University of Louvain; and Verheyen, in his turn, had copied them from a monumental work "Vivae Imagines Partium Corporis humani" by André Vésale, or Andreas Vesalius, a famous Belgian anatomist born in this city of Brussels in 1514. The Japanese text "Kaitai Shinsho" contains exact reproductions of pictures drawn by Vesalius, even with the typographical mark of the well-known Antwerp publisher "Plantin". Thus, it is through the intermediary of Kulmus and Verheyen that the Japanese got to know Vesalius, the founder of modern anatomical studies, and it is by the Low Countries—made up of Belgium as well as Holland at that time—that Japan was initiated into modern medicine and surgery.

Today, medical science carries immense possibilities for the future of mankind. It is capable of increasing still further human welfare, thus bringing man closer to God. But it is also capable of producing demoniac creatures such as Frankenstein monsters. The theme of this year's conference, "Personality and Neurosurgery" seems to me extremely interesting and fascinating as indeed have been the previous topics, "pain" and "plasticity in the nervous system". Is it really possible to change and improve personality by means of modern neurosurgery? If so, what a marvel! But at the same time as awe and admiration, one cannot help feeling great dread, when one imagines that the world of Dr. Jekyll and Mr. Hyde is no more that of romantic fiction, but is potentially here. I am certain we are all agreed that medical science must never be allowed to tamper with human personality in order to promote evil.

As your esteemed Academy bears the titel "Eurasian", Mr. President, you have assembled so many distinguished scholars and men of science from East as well as West; let me congratulate you and your staff on this remarkable achievement. For us professional diplomats, as you know, East and West have a special connotation, particularly at this moment. When we say "East-West relations", we immediately think of the Reagan-Gorbachev meeting, INF negotiations, SS-20 and Pershing I or II, Star Wars and the like. In this context, East and West are taken as political entities, and East, more often than not, presents itself as an adversary to the West which, curiously enough, includes a Far-Eastern country like Japan. Relations between these two blocs or groups are sometimes strained and antagonistic. In this regard, the famous lines of Kipling "East is East and West is West, and never the twain shall meet" may perhaps sound right.

I believe, however, that here in this forum, differences in social and political systems play no significant role, and geographical East meets West in a very friendly and constructive atmosphere, despite undeniable differences of race, culture and religion. Kipling's other lines must not be forgotten: "There is neither East nor West, Border nor Breed, nor Birth, when two strong men stand face to face, though they come from the ends of the earth".

Ladies and Gentlemen,

You are the strong men of our time. Whether you stand face to face in loyal competition or side by side in the common effort, I wish you all, every success both for your Convention here and for your work in the future.

Thank you.

Acta Neurochirurgica, Suppl. 44, 5 (1988)

Address

by His Excellency, Mr. **Willy de Clercq**

Commissioner at the European Community

1. If geographiccaly speaking Europe and Asia are two distinct continents, they have always been worlds apart in terms of mentality and culture. It is striking how, in the course of history, these two worlds have been alternately attracted and repulsed by each other.

It is hardly surprising that the two worlds have a certain historical fear of each other. There was a time when the hordes of Genghis Khan threatened Europe, but also a time when the Europeans adopted a kind of "gunboat diplomacy" in Asia;

There is, on the other hand, a certain attraction between the two worlds. Marco Polo's journey is one example so that in the course of history both worlds have been to have a look at each other and they have integrated certain aspects of the "other" culture into their own.

2. I am pleased to note that the second half of this century has been marked by a significant cultural rapprochement between Europe and Asia. The great progress made in transport, telecommunications and information systems has helped East and West get to know and appreciate each other better each day.

Moreover, the fact that a number of Asian countries have managed to industrialize in a record time has placed the two partners psychologically in a position of equality.

3. In my view there are great opportunities for the world as a whole to make faster technological progress if Europe and Asia can pool their resources. Each culture has its strong side which perhaps complements the other very well. Why then should there not be a partnership between East and West in the interests of all people throughout the world?

4. In my opinion the Academia Eurasiana is an excellent example of how such cooperation between East and West can and must be put into practice. You might perhaps think it is a closed, elitist club which, like every academy, lives a life of its own. However, it is my belief that new ideas are born more easily in the calm of a closed environment. That is why this kind of academy has already existed for several hundred years. Big congresses are excellent places for meeting people, but probably not the most efficient way of producing intellectual cross-pollination, and that is what we are talking about here, for today's convention is least of all a social event. I also believe that if the Academia Eurasiana did not exist, it would have to be invented. It meets a real need.

5. It merely remains for me to wish all convention participants much success, and to hope that in its limited field the Academy can take a step forward towards greater cooperation between East and West.

II. Meaning of Human Personality

Acta Neurochirurgica, Suppl. 44, 9–18 (1988)

Behaviourism East and West

M. Brandeleer

Licentiate in Philosophy, Université Catholique de Louvain, Belgium

In relating Zen Buddhism to psychoanalysis, one discusses two systems, both dealing with theory concerned with the nature of man and with a practice leading to his well-being. Each is a characteristic expression of Eastern and Western thought, respectively.

Zen Buddhism is a blending of Indian rationality and abstraction with Chinese concreteness and realism. Psychoanalysis is as exquisitely Western as Zen is Eastern. Much further back, Greek wisdom and Hebrew ethics are the spiritual godfathers of this scientific – therapeutic approach to man. Psychoanalysis is a scientific method, non religious to its core. Zen is a theory and technique to a chieve "enlightenment", an experience which in the West would be called religious or mystical. Psychoanalysis is a therapy for mental illness; Zen a way to spiritual salvation. Can thus the discussion of the relationship between psychoanalysis and Zen Buddhism result in anything but the statement that there exists no relationship except that of a radical and unbridgeable difference?

What I will attempt to do is to give an answer to these questions. I will not try to give a systematic presentation of Zen Buddhist thought, a task which would transcend my knowledge and experience, nor will I try to give a full presentation of psychoanalysis, which would go beyond the scope of this speech. Nevertheless, I shall in the first part present in some detail those aspects of psychoanalysis which are of immediate relevance to the relation between psychoanalysis and Zen Buddhism and which, at the same time, represent the basic concept of that continuation of Freudian analysis which I would like to call "humanistic psychoanalysis".

As a first approach to our topic, we must consider the spiritual crisis which Western man is undergoing in this crucial historical epoch. It is the crisis which has been described as "malaise", "ennui", "mal du siècle", the deadening of life, the automatization of man, his alienation from himself, from his fellow man and from nature. Man has followed rationalism to the point where rationalism has transformed itself into utter irrationality. Since Descartes, man has increasingly split thought from affect. Control by the intellect over nature, and the production of more and more things, became the paramount aims of life. In this process man has transformed himself into a thing, life has become subordinated to property, "to be" is dominated by "to have".

Where the roots of Western culture, both Greek and Hebrew, considered the aim of life the perfection of man, modern man is concerned with the perfection of things, and the knowledge of how to make them. Western man often is in a state of schizoid inability to experience affect, hence he is anxious, depressed, and desperate.

The abandonment of theistic ideas in the nineteenth century was – seen from one viewpoint – no small achievement. Man took a big plunge incto objectivity. The earth ceased to be the centre of the universe; man lost his central role as the creature destined by God to dominate all other creatures. Studying man's hidden motivation with a new objectivity, Freud reorganized that the faith in an all-powerful, omniscient God, had its root in the helplessness of human existence and in man's attempt to cope with it. He saw that only man can save himself; the teaching of the great teachers, the loving help of parents, friends, can help him, but can help him to dare to accept the challenge of existence and to react to it with all his might and all his heart.

The East, however, was not burdened with the concept of a transcendent father – savior in which the monotheistic religions expressed their longings. Taoism and Buddhism had a rationality and realism superior to that of the Western religion. They could see more realistically and objectively, having nobody but the "awakened" one to guide him, and being able to be guided because each man has within himself the capacity to awake and be enlightened. This is precisely the reason why Eastern religious thought, Taoism and Buddhism – and their blending in Zen Buddhism – assume such importance for the West today. Zen Buddhism helps man to find an answer to the question of his existence, an answer which is essentially the same as that given in the Judaic-Christian tradition, and yet which does not contradict the rationality, realism and independence which are modern man's precious achievements. Paradoxically, Eastern religious thought turns out to be more congenial to Western rational thought than does Western religious thought itself.

Psychoanalysis is a characteristic expression of Western man's spiritual crisis, and an attempt to find a solution. This is explicitly so in the more recent development of psychoanalysis, in "humanist" or "existentialist" analysis. However, before I discuss my own "humanist" concept, I want to show that, quite contrary to a widely held assumption, Freud's own system transcended the concept of "illness" and "cure" and was concerned with the "salvation" of man. If we look more closely, we find that behind this concept of a medical therapy was as entirely different interest, rarely expressed by Freud, and probably rarely conscious even to himself.

What was Freud's vision for man's future? Freud answered this question perhaps most clearly in the sentence: "Where there was Id – there shall be Ego". His aim was the domination of irrational and unconscious passions by reason. Freud's aim was the optimum knowledge of truth, and that is the knowledge of reality; this knowledge to him was the only guiding light man had on this earth.

While Freud represents the culmination of Western rationalism, it was his genius to overcome at the same time the false rationalistic and superficially optimistic aspect of rationalism and to create a synthesis with romanticism, the very movement which during the nineteenth century opposed rationalism by its own interest in and reverence for the irrational, affective side of man. With regard to the treatment of the individual, Freud was also more concerned with a philosophical and ethical aim than he was generally believed to be.

Those familiar with Eastern thought, and especially with Zen Buddhism, will notice that the factors which I am going to mention are not without relation to concepts and thoughts of the Eastern mind. The principle to be mentioned here first is the concept that knowledge leads to transformation, that theory and practice must not be separated, that in the very act of knowing oneself, one transforms oneself. In still another aspect Freud's method has a close connection with Eastern thought in the fact that he did not share the high evaluation of our conscious thought system, so characteristic of modern Western man.

In his wish to arrive at insight into the real nature of a person, Freud wanted to break through the conscious thought system, by his method of free association. Free association was to by-pass logical, conscious, conventional thought. It was to lead into a new source of our personality, namely, the unconscious. Whatever criticism may be made of the contents of Freud's unconscious the fact remains that by emphasizing free association as against logical thought he moved in a direction which had been developed much farther and much more radically in the thought of the East.

There is one further point in which Freud differs radically from the contemporary Western attitude. I refer here to the fact that he was willing to analyze a person for one, two, three or even more years. One would rather say that the time spent in such a prolonged analysis is not worth while, if we consider the social effect of a change in one person. Freud's method makes sense only if one transcends the modern concept of "value", of the proper relationship between means and ends, if one takes the position that one human being is not commensurable with any thing, that his emancipation, his well-being his enlightenment, or whatever term we might want to use, is a matter of "ultimate concern" in itself, then no amount of time and money can be related to this aim in quantitative terms.

The foregoing remarks are not meant to imply that Freud in his conscious intentions was close to Eastern thought. He was much too much a son of eighteenth and nineteenth century thought to be closer to Eastern thought as expressed in Zen Buddhism. He saw man as fundamentally egoistical, and related to others only by mutual necessity of satisfying instinctual desires. Pleasure for him was relief of tension, not the experience of joy. Brotherly love was an unreasonable demand contrary to reality, mystical experience a regression to infantile narcissism.

What I have tried to show is that in spite of obvious contradictions to Zen Buddhism, there were nevertheless elements in Freud's system which transcended the conventional concepts of illness and cure, elements which led to a further development of psychoanalysis which has a more direct and positive affinity with Zen Buddhist thought.

Before we come to the discussion of the connection between this "humanistic" psychoanalysis and Zen Buddhism, I want to point to the change in the kind of patients who come for analysis, and the problem they present. At the beginning of this century the people who came to the psychiatrist were mainly people who suffered from symptoms. In other words, they were sick in the sense in which the word "sickness" is used in medicine, and their concept of "wellness" was not to be sick.

The new "patients" who function socially, are not sick in the conventional sense, but do suffer from the "inner deadness" I have been discussing above. These new patients come to the psychoanalyst without knowing what they really suffer from. They complain about being depressed, being unhappy, not enjoying life. However, these patients usually do not see that their problem is not that of depression ... These various complaints are only the conscious form in which our culture permits them to express something which lies much deeper. The common suffering is the alienation from oneself, from one's fellow man and from nature; the awareness that life runs out of one's hand like sand and that one will die without having lived. For those who suffer from alienation, cure does not consist in the absence of illness, but in the presence of well-being.

However, if we are to define well-being, we meet with considerable difficulties. Any attempt to give a tentative answer to the problem of well-being must transcend the Freudian form of reference and lead to a discussion, incomplete as it must be, of the basic concept of human existence.

Only in this way can we lay the foundation for the comparison between psychoanalysis and Zen Buddhist thought.

The first approach to a definition of well-being can be stated thus: to be in accord with the nature of man. If we go beyond this formal statement, the question arises: what is being, in accordance with the conditions of human existence? What are these conditions?

Human existence poses a question. Man is thrown into this world without his volition, and taken away from it again without his volition. In contrast to the animal, which in its instincts has a "built-in" mechanism of adaptation to its environment, man lacks this instinctive mechanism. He has to live his life he is not lived by it. He is in nature, yet he transcends nature. At the moment of birth, life asks man a question, and this question he must answer. He must answer it at every moment; not his mind, not his body, but he, the whole man must answer it. The question is always the same. However, there are several answers, or basically, there are two answers.

One answer is to overcome separateness and find unity by regression to the state of unity which existed before awareness ever arose, the other answer is to be fully born, to develop one's awareness, one's reason, one's capacity to love to such a point that one arrives at a new harmony, at a new oneness with the world.

When we speak of birth, we usually refer to the act of physiological birth. But in many ways the significance of this birth is overrated. Birth is not one act, it is a process. The aim of life is to be fully born, though its tragedy is that most of us die before we are thus born. Death occurs when birth stops. Physiologically, our cellular system is in a process of continual birth; psychologically, however, most of us cease to be born at a certain point. The regressive attempt to answer the problem of existence can assure different forms; what is common to all of them is that they necessarily fail and lead to suffering.

These different goals and the ways to attain them are not primarily different systems of thought. They are different ways of being, different answers of the total man to the question which life asks him. They are the same answers which have been given in the various religious systems which made up the history of religion. Religion is the formalized and elaborate answer to man's existence, and since it can be shared in consciousness and by ritual with others, even the lowest religion creates a feeling of reasonableness and of security by the very communion with others. In order to understand any human being, one must know what his answer to the question of existence is, or, to put it differently, what his secret, individual religion is, to which all his efforts and passions are devoted.

Returning now to the question of well-being, how are we going to define it in the light of what has been said so far? Well-being is the state of having arrived at the full development of reason: reason not in the sense of merely intellectual judgement, but in that of grasping truth by "letting things be as they are" (to use Heidegger's term). Well-being is possible only to the degree to which one is open, responsive, sensitive, awake, empty (in the Zen sense).

Although such attempt may be found in individuals of relatively primitive societies, the grat dividing line for the whole of humanity seems to lie in the period between roughly 2,000 B.C. and the beginning of our era. Taoism and Buddhism in the Far East, Ikhnaton's religious revolutions in Egypt, the Zoroastrian religion in Persia, the Moses religion in Palestine. Unity in sought in all these religions. This new unity has as a premise the full development of man's reason, leading to a stage in which reason no longer separates man from his immediate, intuitive grasp of reality. There are many new symbols for the new goal which lies abhead, and not in the past: Tao, Nirvana Enlightenment, the Good, God.

What is common to Jewish-Christian and Zen Buddhist thinking is the awareness that I must give up my "will" (in the sense of my desire to force, direct, strangle the world outside of me and within me). In Zen terminology this is often called "to make oneself empty", which does not mean something negative, but means the openness to receive. In Christian terminology this is often called "to slay oneself and to accept the will of God". As far as the popular interpretation and experience is concerned, this formulation means that instead of making decisions himself, man leaves the decisions to an omniscient, omnipotent father; it is clear that in this experience man does not become open and responsive, but obedient and submissive. Paradoxically, I truly follow God's will if I forget about God.

In the foregoing part I have tried to outline the ideas of man and of human existence which underlie the goals of humanistic psychoanalysis. We must now proceed to describe the specific approach through which psychoanalysis tries to accomplish its goal.

The most characteristic element in it is without any doubt its attempt to make the unconscious conscious. Questions immediately arise: what is the unconscious? What is consciousness? How does the unconscious become conscious? And if this happens, what effect does it have? First of all we must consider that the terms conscious and unconscious are used with several different meanings. In one meaning, which might be called functional, "conscious" and "unconscious" refer to a subjective state within the individual. Saying that he is conscious of this or that psychic content means that he is aware of affects, of desire, of judgement etc ... Unconscious, used in the same sense, refers to a state of mind in which the person is not aware of his inner experience. We must remember that "unconscious" does not refer to the absence of any impulse, feeling, desire, fear ... but only to the absence of awareness of these impulses.

Quite different from the use in the functional sense just described is another use in which one refers to certain localities in the person and to certain contents connected with these localities. Here "the conscious" is one part of the personality with specific contents, and the "unconscious" is another part of the personality, with other specific contents.

There exists still another use of "conscious" which sometimes leads to confusion. Consciousness is identified with reflecting intellect, the unconscious with unreflected experience. This use does not seem fortunate, intellectual reflection is, of course, always conscious, but not all that is conscious is intellectual reflection. If I look at a person, I am aware of the person, I am aware of whatever happens in me in relation to the person, but only if I have separated myself from him in a subject-object distance, is the consciousness identical with intellectual reflection.

Having decided to speak of unconscious and conscious as states of awareness and unawareness, we must now consider the question of what prevents an experience from reaching our awareness, that is, from becoming conscious.

Why should we be striving to broaden the domain of consciousness, unless this were so? Yet it is quite obvious that consciousness as such has no particular value; in fact, most of what people have in their conscious minds is fiction and delusion; this not so much because people would be incapable of seeing the truth as because of the function of society.

In its historical development each society becomes caught in its own need to survive in the particular form in which it has developed, and it usually accomplishes this survival by ignoring the wider human aims which are common to all man. This contradiction between the social and the universal aim leads also to the fabrication (on a social scale) of all sorts of fictions and illusions which have the function to deny and to rationalize the dichotomy between the goals of humanity and those of a given society.

Right now I want only to emphasize that most of what is in our consciousness is "false consciousness" and that is essentially society that fills us with these fictitious and unreal notions. But the effect of society is not only to funnel fictions into our consciousness, also to prevent the awareness of reality.

The further elaboration of this point leads us straight into the central poblem of how repression or unconsciousness occurs.

The animal has a consciousness of the things around it which we may call "simple consciousness".

Man's brain structure, being larger and more complex, transcends this simple consciousness and is the basis of self consciousness, awareness of himself as the subject of his experience. But perhaps because of its enormous complexity human awareness is organized in various possible ways, and for any experience to come into awareness, it must be comprehensible in the categories in which conscious thought is organized. Some of the categories, such as time and space, may be universal, and may constitute categories of perception common to all men. Others, such as causality, may be a valid category for many, but not for all forms of human conscious perception. This system is in itself a result of social evolution. Every society, by its own practice of living and by the mode of relatedness of feeling and perceiving, develops a system of categories which determines the forms of awareness. This system works, as it were, like a socially conditioned filter.

The question then, is to understand more concretely how this "social filter" operates, and how it happens that it permits certain experience to be filtered through, while others are stopped from entering awareness.

First of all, we must consider that many experiences do not lead themselves easily to being perceived in awareness. Pain is perhaps the physical experience which best lends itself to being consciously perceived; sexual drive, hunger etc. also are easily perceived. However when it comes to a more subtle or complex experience, like seeing a rosebud in the early morning, a drop of dew on it, while the air is still chilly, the sun coming up, a bird singing, this is an experience which in some cultures easily lends itself to awareness (for instance in Japan), while in modern Western culture this same experience will usually not come into awareness because it is not sufficiently "eventful" to be noticed!

There are many affective experiences for which a given language has no word, while another language may be rich in words which express these feelings. Generally speaking, it may be said that an experience rarely comes into awareness for which the language has no words! The whole language contains an attitude of life, is a frozen expression of experiencing life in a certain way.

Here are a few examples. There are languages in which the verb form "it rains", for instance, is conjugated differently depending on whether I say that it rains because I have been out in the rain and have got wet, or because somebody has told me that it is raining. It is quite obvious that the emphasis of the language on these different sources of experiencing a fact has a deep influence on the way people experience facts. Or, in Hebrew the main principle of conjugation is to determine whether an activity is complete (perfect) or incomplete (imperfect), while the time in which it occurs – past, present, future – it expressed only in a secondary fashion. In Latin, both principles (time and perfection) are used together.

Language by its words, its grammar, its syntax, by the whole spirit which is frozen in it, determines how we experience, and which experiences penetrate to our awareness.

The second aspect of the filter which makes awareness possible is the logic which directs the thinking of people in a given culture. A good example of this is the difference between Aristotelian and paradoxical logic. Aristotle stated: "It is impossible for the same thing at the same time to belong and not to belong to the same thing and in the same respect". In opposition to Aristotelian logic is what we might call paradoxical logic which assumes that A and non-A do not exclude each other as predicates of X. Paradoxical logic was predominant in Chinese and Indian thinking, in Heraclitus philosophy, and then again under the name of dialectics in the thought of Hegel. The general principle of paradoxical logic has been clearly described in general terms by LAO-TSE: "Words that are strictly true seem to be paradoxical" and by CHUANG TZU: "That which is one is one. That which is not-one is also one". Inasmuch as a person lives in a culture in which the correctness of Aristotelian logic is not doubted, it is exceedingly difficult for him to be aware of experiences which contradict Aristotelian logic, which are hence nonsensical from the standpoint of his culture.

The third aspect of the filter, apart from language and logic, is the content of experiences.

Every society excludes certain thoughts and feelings from being thought felt, and expressed. There are things which are not only "not done" but which are even "not thought". In stating the thesis that contents which are incompatible with socially permissible ones are not permitted to enter the realm of awareness, we raise two further questions. Why are certain contents incompatible with a given society?

Furthermore, why is the individual so afraid of being aware of such forbidden contents?

As to the first question, I must refer to the concept of the "social character". Any society, in order to survive, must mould the character of its members in such a way that they want to do what they have to do; their social function must become internalized and transformed into something they feel driven to do, rather than something they are obliged to do. Societies, of course, differ in the rigidity with which they enforce their social character, and the observation of the taboos for protecting this character, but in all societies there are taboos, the violation of which results in ostracism.

We come, then, to the conclusion that consciousness and unconsciousness are socially conditioned. I am aware of all my feelings and thoughts which are permitted to penetrate the threefold filter of language, logic and social character. Experiences which cannot be filtered through remain outside of awareness, that is, they remain unconscious. The more a society approximates the human norm of living, the less is there a conflict between isolation from society and from humanity. The greater the conflict between social aims and human aims, the more is the individual torn between the two dangerous poles of isolation.

As to the contents of the unconscious, no generalization is possible, but one statement can be made: it always represents the whole man, with all his potentialities for darkness and light; it always contains the basis for the different answers which man is capable of giving to the question which existence poses. The content of the unconscious, then, is neither the good nor the evil, the rational nor the irrational; it is both; it is all that is human. The unconscious is the whole man – minus that part of man which corresponds to his society. Consciousness represents social man, the accidental limitations set by the historical situation into which an individual is thrown. Unconsciousness represents universal man, the whole man, rooted in the Cosmos.

Defining consciousness and unconsciousness as we have done, what does it mean if we speak of making the unconscious conscious?

When we free ourselves from the limited concept of Freud's unconscious and follow the concept presented above, then Freud's aim gains a wider and more profound meaning. Making the unconscious conscious transforms the mere idea of the universality of man into the living experience of this universality, it is the experiential realization of humanism.

Taking into account what has been said above about the stultifying influence of society and considering our wider concept of what constitutes unconsciousness, we may begin by saying that the average person, while he thinks he is aware, is actually half asleep. By "half asleep" I mean that his contact with reality is a very partial one, he is aware of reality only to the degree of which his social functioning makes it necessary, he is aware of material and social reality inasmuch as he needs to be aware of it in order to manipulate it. We can thus differentiate between what a person is conscious of, and what he becomes conscious of.

What happens then in the process in which the unconscious becomes conscious? In answering this question we had better reformulate it. Our question then should rather be: what happens when I become aware of what I have not been aware of before? Could this be the same experience Zen Buddhists call "enlightenment"?

While I shall return later to this ultimate question, I want at this point to discuss further a crucial point namely, the nature of insight and knowledge which is to affect the transformation of unconsciousness into consciousness. Doubtlessly, in the first years of his psychoanalytic research, Freud shared the conventional rationalistic belief that knowledge was merely intellectual, theoretical knowledge. This intellectual knowledge, called "interpretation", was supposed to effect a change in the patient. But soon Freud and other analysts had to discover the truth of Spinoza's statement that intellectual knowledge is conducive to change only inasmuch as it is also affective knowledge. As long as the patient remains in the attitude of the detached scientific observer, taking himself as the object of his investigation, he is not in touch with his unconscious, except by thinking about it, he does not experience the wider, deeper reality within himself. This does not mean that thinking and speculation may not precede the act of discovery, but the act itself is always a total experience. It is total in the sense that the whole person experiences it, it is an experience which is characterized by its spontaneity and suddenness. The importance of this kind of experiential knowledge lies in the fact that it transcends the kind of knowledge and awareness in which the subject-intellect observes himself as an object, and thus that it transcends the Western, rationalistic concept of knowing. The importance of this kind of experience for the problem of Zen Buddhism will be clarified later, in the following part dealing with the discussion of Zen.

So far I have discussed man's existence and the question it poses; the nature of well-being defined as the overcoming of alienation and separateness; the specific method by which psychoanalysis tries to attain its goal, namely, the penetration of the unconscious. I have dealt with the question of what the nature of unconsciousness and of consciousness is; and what "knowing" and "awareness" mean in psychoanalysis.

In order to prepare the ground for a discussion of the relationship between psychoanalysis and Zen, I must speak of those principles of Zen which have an immediate bearing on psychoanalysis. The essence of Zen is the acquisition of enlightenment (Satori). One who has not had this experience can never fully understand Zen. Since I have not experienced satori, I can talk about Zen only in a tangential way, and not as it ought to be talked about – out of the fullness of experience. As far as this goes, Zen is not more difficult for the European than Herclitus, Hegel, or Heidegger. The difficulty lies in the tremendous effort which is required to acquire satori; this effort is more than most people are willing to undertake, and that is why satori is rare even in Japan.

Even though I cannot talk of Zen with any authority, the good fortune of having read Dr. Suzuki's books, heard quite a few lectures, has given me at least an approximate idea of what constitutes Zen, an idea which I hope enables me to make a tentative comparison between Zen Buddhism and psychoanalysis.

What is the basic aim of Zen? To put it in Suzuki's words: "Zen in its essence is the art of seeing into the nature of one's being, and it points the way from bondage to freedom ..." We can say that Zen liberates all the energies properly and naturally stores in each of us, which are in ordinary circumstances cramped and distorted so that they find no adequate channel for activity. Generally, we are blind to this fact, that we are in possession of all the necessary faculties that will make us happy and loving towards one another. We find in this definition a number of essential aspects of Zen which I should like to emphasize: Zen is the art of seeing into the nature of one's being, it is a way from bondage to freedom; it liberates our natural energies and it impels us to express our faculty for happiness and love.

The final aim of Zen is the experience of enlightenment, called satori. I would like to stress some aspects which are of special importance for us, Westerners. Satori is not an abnormal state of mind, it is not a trance in which reality diappears, as it can be seen in some religious manifestations. As Jōshū declared "Zen is your every day thought, it all depends on the adjustment of the hinge, whether the door opens in or opens out". It is quite clear that satori is the true fulfillment of the state of well-being. If we would try to express enlightenment in psychological terms, I would say that it is a state in which the person is completely turned to the reality outside and inside of him. He is aware of it – that is, not his brain, nor any other part of his organism, but he, the whole man. He is aware of it, not as of an object over there which he grasps with this thought, but it, the flower, the man, in its, or his, full reality.

To be enlightened means "the full awakening of the total personality to reality". That means not to relate oneself to the world receptively, exploitatively, or in the marketing fashion, but creatively, actively (in Spinoza's sense). In the state of full productiveness there are no veils which separate me from the "not me". The object is not an object any more; it does not stand against me, but is with me.

The state of productiveness is at the same time the state of highest objectivity, I see the object without distortions by my greed and fear.

I see it as it or he is, not as I wish it or him to be or not to be. In this mode of perception there are no parataxic distortions. Satori appears mysterious only to the person who is not aware to what degree his perception of the world is purely mental, or parataxical.

That the undistorted and non-cerebral perception of reality is an essential element of Zen experience is expressed quite clearly in one Zen story. It is the story of a master's conversation with a monk:

"Do you ever make an effort to get disciplined in the truth?"

"Yes, I do."

"How do you exercise yourself?"

"When I am hungry, I eat; when I am tired, I sleep."

"This is what everybody does; can they be said to be exercising themselves in the same way as you do?"

"No."

"Why not?"

"Because when they eat, they do not eat, but are thinking of various other things, thereby allowing themselves to be disturbed; when they sleep they do not sleep, but dream of a thousand and one things."

The story hardly needs any explanation. The average person, driven by insecurity, greed, fear, is constantly emeshed in a world of phantasies (not necessarily being aware of it), in which he clothes the world in qualities which he projects into it, but which are not there. This was true at the period when this conversation took place; how much more is it true today, when almost everybody sees, hears, feels, and tastes with his thoughts, rather than with those powers within himself which can see, hear, feel and taste.

Zen is aimed at the knowledge of one's own nature. It searches to "know oneself", but this knowledge is not the "scientific" knowledge, the knowledge of the knower – intellect who knows himself as object. As Suzuki has put it: "The basic idea of Zen, is to come into touch with the inner working of one's being, and to do this in the most direct way possible, without resorting to anything external or super-added."

This difference between intellectual and experiential

knowledge is of central importance for Zen, and, at the same time, constitutes one of the basic difficulties the Westerner has in trying to understand Zen. The West, for two thousand years (and with only few exceptions, such as the mystics) has believed that a final answer to the problem of existence can be given in thought; the "right answer" in religion and in philosophy is of paramount importance. By this insistence the way was prepared for the fluorishing of the natural sciences.

Here the right thought, while not giving a final answer to the problem of existence, is inherent in the method and necessary for the application of the thought to practice, that is, for technique. Zen, on the other hand, is based on the premise that the ultimate answer to life can not be given in thought. "The intellectual groove of "yes" and "no" is quite accommodating when things run their regular course; but as soon as the ultimate question of life comes up, the intellect fails to answer satisfactorily." For this very reason, the experience of satori can never be conveyed intellectually. It is an experience which no amount of explanation and argument can make communicable to others, unless the latter themselves had it previously. For a satori turned into a concept ceases to be itself; and there will no more be a Zen experience.

As a further consequence, the concept of participation or empathy is unacceptable to Zen thought. "The idea of participation or empathy is an intellectual interpretation of primary experience, while as far as the experience itself is concerned, there is no room for any short of dichotomy." The intellect, however, obtrudes itself and breaks up the experience in order to make it amenable to intellectual treatment, which means a discrimination or bifurcation. The original feeling of identity is then lost and intellect is allowed to have its characteristic way of breaking up reality into pieces. Not only intellect, but any authoritative concept or figure, restricts the spontaneity of experience; thus Zen does not attach any intrinsic importance to the sacred sutras or to their exegesis by the wise and learned. Personal experience is strongly acting against authority and objective revelation. In Zen God is neither denied nor insisted upon. Zen wants absolute freedom, even from God.

In accordance with Zen's attitude towards intellectual insight, its aim of teaching is not as in the West an ever-increasing subtlety of logical thinking, but its method consists in putting one in a dilemma, out of which one must contrive to escape not through logic indeed but through a mind of higher order. The attitude of the Zen master to his student is bewildering to the modern Westerner who is caught in the alternative between an irrational authority which limits freedom and exploits its object, and a laissez-faire absence of any authority. Zen represent another form of authority. The master does not call the student, he wants nothing from him, not even that he becomes enlightened; the student comes of his own free will, and he goes of his own free will. But inasmuch as he wants to learn from the master, that is, that the master knows what the student wants to know, and does not yet know. The Zen master is characterized at the same time by the complete lack of irrational authority and by the equally strong affirmation of that undemanding authority, the source of which is genuine experience.

Zen cannot possibly be understood unless one takes into consideration the idea that the accomplishment of true insight is indissolubly connected with a change in character. Here Zen is rooted in Buddhist thinking, for which characterological transformation is a condition for salvation. The attitude towards the past is one of gratitude, towards the present of service, and towards the future of responsibility. To live in Zen "means to treat yourself and the world in the most appreciative and reverential frame of mind", and attitude which is the basis of "secret virtue", a very characteristic feature of Zen discipline.

What follows from our discussion of psychoanalysis and Zen as to the relationship between the two?

You may have been struck by now by the fact that the assumption of incompatibility between Zen Buddhism and psychoanalysis results from a superficial view of both. Quite to the contrary, the affinity between both seems to be much more striking. What follows now is devoted to a detailed elucidation of this affinity.

The description of Zen's aim could be applied without change as a description of what psychoanalysis aspires to achieve; insight into one's own nature, the achievement of freedom, happiness and love, liberation of energy, salvation from being insane or crippled.

Before we arrive at the central issue of the connection between both, I wish to consider some more peripheral affinities. First to be mentioned is the ethical orientation common to Zen and to psychoanalysis. Yet neither of them is primarily an ethical system. The aim of Zen transcends the goal of ethical behaviour, and so does psychoanalysis. It might be said that both systems assume that the achievement of their aim brings with it an ethical transformation. They do not tend to make a man lead a virtuous life by the suppression of the "evil" desire, but they expect that the evil desire will melt away and disappear under the light and warmth of enlarged consciousness.

Another element common both systems is their insistence on independence from any kind of authority. Yet it might be asked, does this anti-authoritarian attitude not contradict the significance of the person of the master in Zen, and of the analyst in psychoanalysis? Again, this question points to an element in which there is a profound connection between both. In both systems a guide is needed, one who has himself gone through the experience the patient (student) under his care is to achieve. The master is willing to guide him, but only under one condition: that the student understands that, much as the master wants to help him, the student must look after himself. None of us can save anybody else's soul. One can only save onseself. Related to the attitude of the analyst is another affinity between Zen and psychoanalysis. The "teaching" method of Zen is to drive the student into a corner, as it were. The "Koan" makes it impossible for the student to seek refuge in intellectual thought, the koan is like a barrier which makes further flight impossible. The analyst does or should do something similar. He must avoid the error of feeding the patient with interpretations and explanations which only prevent the patient from making the jump from thinking into experiencing.

In the process of making the unconscious conscious, of arriving at the full and hence unreflected reality of experience, both the conscious and the unconscious must be trained. The conscious must be trained to loosen its reliance on the conventional "filter", while the unconscious must be trained to emerge from its secret, separate existence, into the light. But in reality, speaking of the training of consciousness and unconsciousness means using metaphors. Man must be trained to drop his repressedness and to experience reality fully.

Dr. Suzuki suggests calling this unconscious the Cosmic Unconscious.

The aim of the full recovery of unconsciousness by consciousness is quite obviously much more radical than the general psychoanalytic aim. The reasons for this are easy to see. To achieve this total aim requires an effort far beyond the effor most persons are willing to make. But quite apart from this question of effort, even the visualization of this aim is possible only under certain conditions. First of all, this radical aim can be envisaged only from the point of view of a certain philosophical position. There is no need to describe this position in detail. Suffice it to say that it is one in which well-being is conceived in terms of full union, the immediate and uncontaminated grasp of the world. This aim could not be better described than has been done by Suzuki in terms of "the art of living". One must keep in mind that any such concept as the art living grows from the soil of a spiritual humanistic orientation, as it underlies the teaching of Buddha and of the prophets. Unless it is seen in this context, the concept of the art of living loses all that is specific, and deteriorates into a concept that goes under the name of "happiness".

In stating all this, however, we must be prepared to be confronted with an objection. If, as I said, the achievement of the full consciousness of the unconscious is as radical and difficult an aim as enlightenment, does it make any sense to discuss this radical aim as something which has any general application?

If there were only the alternative between full enlightenment and nothing, then indeed this objection would be valid. But this is not so. In Zen there are many stages of enlightenment, of which satori is the ultimate and decisive step. But, as far as I understand, value is set on experiences, which are steps in the direction of satori, although satori may never be reached. Dr. Suzuki illustrated this point in the following way: if one candle is brought into an absolutely dark room, the darkness disappears, and there is light. But if ten or a hundred or a thousand candles are added, the room will become brighter and brighter. Yet the decisive change was brought about by the first candle which penetrated the darkness.

So far we have spoken about aims. But as to the methods of achieving these aims, psychoanalysis and Zen are, indeed, quite different. The method of Zen is, one might say, that of a frontal attack on the alienated way of perception by means of the "sitting", the koan. Of course, all this is not a "technique" which can be isolated from the premise of Buddhist thinking.

The psychoanalytic method is entirely different from the Zen method. It trains consciousness to get hold of the unconscious in a different way. It directs attention to that perception which is distorted; it leads to a recognition of the fiction within oneself, it widens the range of human experience by lifting repressedness. The analytic method is psychological-empirical.

What can be said with certainty is that the knowledge of Zen, and a concern with it, can have a most fertile and clarifying influence on the theory and technique of psychoanalysis. Zen, different as it is in its method can sharpen the focus, throw new light on the nature of insight, and heighten the sense of what it is to see, of what it is to be creative.

If further speculation on the relation between Zen and Psychoanalysis is permissible, one might think of the possibility that psychoanalysis may be significant to the student of Zen. I can visualize it as a help in avoiding the danger of a false enlightenment one which is purely objective. Analytic clarification might help the Zen student to avoid illusions, the absence of which is the very condition of enlightenment.

How could such understanding be possible, were it not for the fact that "Buddha nature is in all of us", that man and existence are universal categories, and that the immediate grasp of reality, waking up, and enlightenment, are universal experiences.

Correspondence: M. Brandeleer, M.D., Licentiate in Philosophy, Université Catholique de Louvain, 14, ave. Docteur Zamenhoff, B-1070 Bruxelles, Belgium.

Acta Neurochirurgica, Suppl. 44, 19–32 (1988)

The Problem of Mind in Eastern Philosophy*

Hajime Nakamura

Prof. Emeritus, University of Tokyo, Director, The Eastern Institute, Inc.

I. The Concept of Mind in Eastern Philosophy

Throughout a long tradition of Eastern philosophy the problem of mind has played an important role, although there has been a variety of implications, differing with peoples and traditions.

The concept of "mind" has been expressed with the words "*manas*" or "citta" in Indian philosophy, whereas in Chinese philosophy the word "hsing" (心) has been used to denote the concept. In Japanese "shin", the Japanized term of Chinese "hsing" has often been used, although the indigenous term "kokoro" is also current in colloquial use.

The Indian term manas is used in the meaning of "mind", "intellect", "perception", "sense", "will", "soul", or "thought" in the Rigveda, which is supposed to be the most ancient literature of the Indo-European peoples. In Avesta and old Persian the same word is used in similar meanings. A certain scholar enumerates seventeen meanings of the Indian word *manas* with minute differentiations[1]. When expressed in German, *manas* means "Sinn", "Geist", "Verstand", "Wahrnehmung", "Empfindung", "Wille", "Seele" or "Gedanke"[2]. This word derived from the verbal form *manyate*, which means *he thinks, er denkt*, and which is etymologially connected with Greek *menō, memora*; Latin *meminisse, monere*; German *meinen*, English *mean*[3]. The Latin and English word mentor is derived from the same etymological origin[4].

On the other hand, the Indian term *citta*, the past participle of the verb *cit/cint*, "to think" mean "thinking", "observing"; "Denken", "Beobachten" or "Verstand" or "Plan", "Absicht" in German[5]. In later days both terms, *manas citta*, were used almost as synonyms, just meaning "mind"[5].

In later Indian philosophy *citta* denotes whatever is experienced or enacted through the mind. So, *citta* comprises, 1) observing, 2) thinking, and 3) desiring or intending; that is to say, the functions of both the reasoning faculty and the heart.

The Chinese word *hsing* originally represents the physical form of heart. The Sanskrit equivalent to heart is *hṛd* or *hṛdaya*. In the Rigveda hṛd (heart) was regarded as the abode of mental functions, feelings or emotions[6]. Hṛd derived from the same etymological origin as Greek *kardia*, Latin *cor* (*cordis*), German *Herz* and English *heart*.

In the early philosophical literature of India mind was the subject of one's existence or something which is most essential to a human being. It is mind that attains spiritual liberation.

"By the mind, indeed, is this [realization] to be attained"[7].

"By the mind alone is It [the ancient, primeval Brahman] to be perceived.

There is on earth no diversity.

He gets death after death,

Who perceives here seeming diversity"[8].

At the same time mind is the organ of thinking, and for that reason it cannot get rid of its own limitations. Discussing Brahman, the absolute, a verse of an Upaniṣad says:

"Not by speech, not by mind,

Not by sight can He be apprehended.

How can He be comprehended.

Otherwise than by one's saying "He ist"?"[9]

In this sense mind is a product of Brahman the absolute:

"From Him is produced breath,

---

* This paper was delivered in Hakone, Japan (2nd Convention of the Academy, 1986). The Academy officers after seeing its content considered it more suitable for inclusion in the present philosophical contributions.

Mind, and all the senses
Space, wind, light, water,
And earth, the supporter of all"[10].

Such as idea was inherited by Yoga practitioners. According to the Yoga philosopohy it is mind that has to perform religious duty. Mind is that where the residua of defilements stay. It is also mind that is emancipated spiritually. Yoga is the suppression of functions of mind[12]. Yoga practioners no longer care to live in a mind which has already performed its duty; their substratum is gone[11].

From a different angle Shao Yung (邵雍, 1011–1077) the Neo-Confucian scholar of China said: "Man is central in the universe, and the mind is central in man ... Therefore, the gentleman highly values the principle of centrality"[13].

Anyhow, Eastern philosophers tend to regard mind as essential to human existence.

References

1. Grassmann H (1976) Wörterbuch zum Rigveda. Otto Harrassowitz, Wiesbaden, col. p 993 f
2. Mayrhofer M (1963) Kurzgefaßtes etymologisches Wörterbuch des Altinischen. Bd 2. Carl Winter, Heidelberg, p 573
3. Monier-Williams Sanskrit English Dictionary. p 738
4. Cf. Mayrhofer op. cit. 2, p 583
5. Grassmann op. cit. col. 451; Mayrhofer (1956) oc. cit. Bd 1, p 387

5.' Yaṃ ca kho etam ... vuccati dittam iti pi mano iti pi viññānam iti pi. (Saṃyutta-Nikāya, PTS edition, vol 2, p 94) cittam mano 'tha vijñāñam ekārthaḥ (Abhidharmakośa II, 34) cinotīti cittam. manute iti manaḥ. vijānātīti vijñānam. cittam śubhāśubhair dhātubhir iti cittam. tad evāśrayabhūtam manaḥ. āśritabhūtaṃ vijñānam ity apare. (Abhidharmakośabhāṣya) p 61, 1.21—p 62, 1.1 Chinese tr. p 21 c). "The menaing of *Citta* best understood when explaining it by expressions familiar to us *with all may heart, heart and soul.*" From the intellectual point of view the term *Citta* may be replaced by Manas or thinking and *Viññāna* or understanding." (Dipak K. Barua (1985) Consciousness or Citta as Revealed in the Early Pāli Texts. The Mahabodhi, vol 93, Number 1–3, Januar–March, pp 9–11). For materials, cf. Pāli Text Society's Pāli-English Dictionary, s.v. citta, mano, viññāṇa

6. Heart H means especially "Herz als Sitz der Empfindungen, namentlich des Wohlwollens, der Freude oder Furcht der Begeisterung, der Andacht, aus dem Lied und Gebet entsprungen." (Grassmann op. cit. col. 1678)
7. Kaṭha-upaniṣad IV, 1, 11
8. Bṛhadāraṇyaka-upaniṣad, IV, 4, 19
9. Kaṭha-up. VI, 12
10. Muṇḍaka-upaniṣad, II, 1, 3
11. Vyāsa's Yogabhāṣya ad IV, 11
12. Yogasūtra, I, 1, 2
13. from *Huang-chi ching-shih* shu, William Theodore de Bary *et al.* (1960) Sources of Chinese Tradition, New York, p 518

II. Mind as the Ethical Basis of Human Existence

Mind is the ethical basis of our conduct. This fact was explicitly emphasized in Early Buddhism.

"(The mental) natures are the result of our mind (manas), are led by our mind, are made up of our mind. If a man speaks or acts with evil mind, sorrow follows him (as a consequnce) even as the wheel follows the foot of the drawer (*i.e.* the ox which draws the cart)[1]".

The purpose of this saying is as follows:

"The mind of man is the only cause of bondage of release; when it is attracted by objects of pleasure it is bound; when it is not attracted by objects of pleasure it is released."

Again the same scripture continues,

"(The mental) natures are the result of our mind (manas), are led by our mind, are made up of our mind. If a man speaks or acts with pure mind, happiness follows him (in consequence) like a shadow that never leaves him"[2].

So we should control our mind.

"Let one be watchful of mind-irritation. Let him practize restraint of mind. Having abandoned the sins of mind let him practize virtue with his mind"[3].

One should control evil thoughts and cultivate good thoughts. Mind is often compared to a jumping monkey. Mind is always changing and changeable.

"The wise who control their body, who likewise control their speech, the wise who control their mind are indeed well controlled"[4].

Evil conduct should be subdued by mind.

"A man should hasten towards the good; he should restraon his thoughts from evil. If a man is slack in doing what is good, his mind (comes to) rejoice in evil"[5].

The mental situation of a religious practitioner is extolled as follows:

"The disciples of Gautama are always well awake; their mind, day and night, delights in abstinence from harm (compassion, love)".

"The disciples of Gautama are always well awake; their mind, day and night, delights in meditation"[6].

Such an attitude was also emphasized in Japan. Referring to education of women, Nake Tōju, the unique Japanese Confucianist, firmly decleared,

"Cultivation of the mind is the essence of all learning"[7].

However, Master Dōgen of Japan, who established a quite unique style of Zen Buddhism, made a revolutionary protest against the general assumption of mind as the agent for Enlightenment. About the relationship between body and mind, he says;

"Is the Way [of liberation] achieved through the mind or through the body? The doctrinal schools speak of the identity of mind and body, and so when they speak of attaining the Way through the body, they explain it in terms of this identity."

In this respect it seems to me that Master Dōgen had in mind the doctrine of Kūkai, Master Kōbō, who asserted the identity of mind and body and the possibility of achieving liberation "in the body" (*i.e.*, in this life). However, he was critical of the traditional opinion. He continues,

"Nevertheless this leaves one uncertain as to what "attainment by the body" truyl means. From the point of view of our school, attainment of the Way is indeed achieved through the body as well as the mind. So long as one hopes to grasp the Truth only through the mind, one will not attain it even in a thousand existences or in aeons of time. Only when one lets go of the mind and ceases to seek an intellectual apprehension of the Truth is liberation attainable. Enlightenment of the mind through the sense of sight and comprehension of the Truth through the sense of hearing are truly bodily attainments. To do away with mental deliberation and cognition, and simply to go on sitting, is the method by which the Way is made an intimate part of our lives. Thus attainment of the Way becomes truly attainment through the body. That is why I put exclusive emphasis upon sitting"[8].

But, in scriptures of Early Buddhism it is taught occasionally in a slightly different way that intelligence or knowledge makes body active and alive, full of vigour, and that if one does not learn, his body becomes descrepit.

A man who does not learn gets old like an ox. "A man who has learnt but little grows like an ox; his flesh increases but his knowledge does not grow"[9].

In spite of man's destiny to get old, his intelligence does not decay. "The splendid chariots of kings wear out; the body also comes to old age, but the virtue of the good never ages, thus the good teach to each other"[10].

Extolling an ideal person, Confucius said; "He is simply a man, who in his eager pursuit of knowledge forgets his food, who in the joy of its attainment forgets his sorrows, and who does not perceive that old age is coming on"[11]. This ideal image has been inherited throughout a long intellectual history of China and Japan. To illustrate, a painter who was so earnest in painting that he did not perceive that "old age was coming on" was highly praised by Tu-po, the famous eighth century Chinese poet, and by others.

A brahmin ascetic asked the Buddha, saying, "I am old, feeble, colourless; my eyes are not clear; my hearing is not good; lest I should perish a fool on the way, tell me the truth of human existence, that I may know how to leave birth and decay (aging) in this world".

The Buddha replied:

"Seeing others afflicted by the body, and seeing heedless people suffer in their bodies, thereafter, shall you be heedful, and leave the body behind, that you may never come to exist in delusion again."

The ascetic asked again:

"Tell me the truth of human existence that I may know how to leave birth and decay (aging) in this world".

The Buddha taught:

"Seeing men seized with desire, tormented and overcome by decay (aging), therefore, you shall be heedful, and leave desire behind, that you may never come to existence of delusion again"[12].

Even in old age one should not lose hope, but the splendour of intelligence radiates. It is said that in this advanced years the Buddha taught as follows:

"There are recluses and Brahmins who say and hold that, as long as a man is in the prime of his youth and early manhood, with a wealth of coalblack hair untouched by grey, and in all the beauty of his prime, so long only are the powers of his mind at their best; but that when he has grown broken and old, aged and stricken in years, and draws to his life's close, then the powers of his mind are in decay. This is not so. I myself am now broken and old, aged and stricken in years and at the close of my life being now round about eighty. Imagine now that I had four disciples—each living to be a full hundred, each of perfect alertness, resolve, and power to reproduce and expound—four disciples as perfect in their scope as a mighty archer of renown, so skilled and dexterous with his bow and so schooled in its use that he can with ease shoot even a feather-weight shaft right over a towering palm.

If you have to carry me about on a litter, yet will my mind still retain its powers.

Of me, it may truly be said that in me a being without delusions has appeared in the world for the welfare and good of many, out of compassion towards the world, for the profit, welfare and good of the people"[13].

In recent years perfectability of mind was discussed in a new setting by Aurobindo Ghosh of India[14].

References

1. Dhammapada, 1
2. ibid. 2

3. ibid. 233
4. ibid. 234
5. ibid. 116
6. ibid. 300, 301
7. Ryūsaku Tsunoda *et al* (1960) Sources of Japanese Tradition. Columbia University Press, New York, p 381
8. Shōbō Genzō Zuimonki (Cited from Tsunoda: op. cit. p 254)
9. Dhammapada 152
10. ibid. 151
11. The Analects, Vii, 18
12. Suttanipāta 1122
13. Majjhima-Nikāya, No. 12 (Mahāsīhanāda-sutta), vol 1, p 82. Translated by Lord Chalmers (1926) Further Dialogues of the Buddha, Sacred Books of Buddhists, vol V, pp 57–58
14. Radhakrishnan S, Moore C (1957) A source book in Indian Philosophy, Princeton University Press, p 580 f

III. The Structure of Mind

A certain Upanishadic thinker admitted a short of hierarchical order among psychic functions or principles in a human existence.

"Speech is more than name ... Mind (manas) is more than speech ... Volition is more than mind. Thought (citta) is more than Volition ...

Meditation is more than thought ...

Understanding (vijñāna) is more than meditation" and so on"[1].

And finally the passage ends with the saying that all these functions or principles arise from the Self (Ātman)[2].

On another occasion mind was regarded as a constituent of one's human existence. When one dies, his body is dissolved and his mind, a constituent of his existence, goes into the moon[3].

The naturalistic philosophy of the Vaiśeṣika school admitted nine substances, and mind was just one of them. "Earth, water, fire, air, ether, time, space, Self (or soul), and mind (are) the only substances"[4]. Here mind and soul were separated.

The characteristic of mind is discussed as follows:

"The appearance and non-appearance of knowledge, on contact of the self with the senses and the objects are the marks (of the existence) of the mind"[5].

The mind of each person is different. So there are many minds according to Praśastapāda[6].

In the philosophy of the Yoga school the object is self-dependent, and common to all the spirits (*puruṣas*). Minds also are self-dependent. They come into relationship with the spirits (*puruṣas*). By their relationship is secured perception which is called enjoyment[7]. But when the yoga practice is brought to perfection, one comes to obtain the knowledge of other minds[8].

Mind perceives objects, but there are some cases of non-perception. The Sāṃkhya school enumerates the cases. Non-perception may be because of extreme distance, extreme proximity, injury to the organs, non-steadiness of the mind, subtlety, veiling, suppression, and blending with what is similar[9].

In the logical system of the Nyāya school, mind was regarded as one of the objects of right knowledge[10], whereas the naturalistic philosophy of the Vaiśeṣika-sūtra denied it.

"Among substances, the self, the mind and others are not objects of perception"[11].

"Substance is the cause of the production of cognition, where attributes and actions are in contact (with the senses)"[12].

Philosophers of ancient India tried to prove the existence of mind. The *Nyāya-sūtra* says:

"The mark of the mind is that there do not arise (in the self) more acts of knowledge than one at a time"[13].

It means that one perceptive cognition occurs on one moment, and another perceptive cognition occurs on the following, next moment. Different perceptions occur at different moments, and they are synthesized by mind with the faculty of apperception. Vātsyāyana, the naturalistic philosopher, comments on this point:

"... Even though at one and the same time several perceptible objects ... are in close proximity to the respective perceptive sense-organs, ... yet there is no simultaneous cognition of them; and from this we infer that there is some other cause [namely, the mind], by whose proximity cognitions appears ... If the proximity of sense-organs to their objects, by themselves, independently of the contact of the mind, were the sole cause of cognitions, then it would be quite possible for several cognitions to appear simultaneously".

Praśastapāda, the Vaiśeṣika philosopher, explains the reasons why we should admit the existence of mind as one substance different from the self or soul.

"Even when there is a proximity of the object to the self and the sense-organ, we find that cognition, pleasure, &c., do not appear, and from this we infer the necessity of an instrumentality other than the aforesaid proximity"[14].

"There is one mind with each body ... If there were many kinds, there would be a multiplicity of the contacts of the self and the mind; and as such we would find the same man having many cognitions and putting forth many efforts at one and the same time. As a matter of fact however we find these appearing only gradually, one after the other; ...

In certain cases we have a notion of more than one cognition appearing at one and the same time; but that is due to the fact of the cognitions following each other very quickly ..."

"The mind, is atomic in its nature ... The mind is atomic in its dimensions; specially as an eternal substance, which is not all-pervading, could not but be atomic, (all intervening dimensions belonging to transient things). The absence of the all-pervading character in the mind is inferred from the nonsimultaneity of congitions; as, if the mind were all-pervading, then it would be in contact with all the sense-organs, at one and the same time; and hence there would be simultaneous cognitions of colour, and taste, &c. ... the absence of all-pervading character in the mind proves its corporeality [or materiality]; ...

This philosopher thought that mind is material; mind consists of matter.

"The materiality of the mind has already been proved by showing that it is atomic. If the mind were the *cogniser*, then, the body would become a common ground of experiences (of all kinds of sensation; or of the mind and the self the two being two distinct cognisers). As a matter of fact, however, we do not find this to be the case, inasmuch as the activity or inactivity of the body is found to follow the purposes of a single agent (or the motive of a single sensation). Hence the mind cannot be regarded as conscious ..."

Mind has qualities of its own.

"The qualities of the mind are—number, dimension, separateness, conjunction, disjunction, priority, posteriority and faculty [tendency or speed]."

The role of mind in the Vaiśeṣika philosophy is instrumentality. Mind plays the role of an instrument of cognition.

An opponent to the Vaiśeṣika school says that consciousness is a quality of mind. But Praśastapāda, the Vaiśeṣika philosopher, refutes this theory, saying

"The mind which you hold to be the substratum of consciousness would be what we call "self" ...

For the following reasong also consciousness cannot belong to the mind; *because the mind is itself of the nature of an instrument*. That is to say, mind is not conscious, because it is an instrument of consciousness, like the jar.

Objection: "The fact of the mind being an instrument has not been accepted (by both parties) as it is held to be the doer or agent".

Reply: If the mind were the doer, then for the perception of pleasure, &c., we should find some other instrument, like the eye for the perception of colour; as no action can ever be produced without an instrument. And if you agree to accept the existence of some such organ, then there would be a difference in name only, as you would also admit the existence of a *doer* (calling it "mind" while we call it "self") and a distinct instrument (calling it something else while we call it "mind")[15].

Mind does not have consciousness[16].

In the philosophy of the Yoga school the function of mind is slightly different. Mind has two aspects, distraction and concentration[17].

According to the dualistic analysis the Sāṃkhya-Yoga school, the spontaneous activities of the mind-stuff, which have to be suppressed before the true nature of the soul or life-monad can be realized, are five:

1. right notions, derived from accurate perception (pramāṇa);
2. erroneous notions, derived from misapprehension (viparyaya);
3. fantasy or fancy (vikalpa);
4. sleep (nidrā); and
5. memory (smṛti).

When these five have been suppressed, the disappearance of desire, and of all other mental activities of an emotional character, automatically follows. The thinking principle, *i.e.*, the mind, assumes the shapes of its perceptions through the functioning of the senses. The formost point of the thinking principle, when meeting objects through the senses, assumes their form. Because of this the process of perception is one of perpetual self-transformation. The mind-stuff is compared, to melted copper, which when poured into a crucible assumes its form precisely. The substance of the mind spontaneously takes on both the shape and the texture of its immediate experience.

Perceptions belong to the sphere of matter. When two material perceptions do not contradict each other, they are regarded as true or right. Nevertheless, even "true" or "right" perceptions are in essence false, and to be suppressed, since they, no less than the "wrong", produce the conception of an "identity of form" (sārūpya) between consciousness-as-mind-stuff and the life-monad (puruṣa or soul).

"According to the analysis of the psyche rendered by the Sāṅkhya school and taken for granted in the disciplines of Yoga, man is "active" (kartar) through the five "organs of action" and "receptive" (bhoktar) through the five "organs of perception". These two sets of five are the vehicles, respectively, of his spontaneity and receptivity. They are known as the "faculties working outward" (bāhyendriya) and function as so many gates and doors, while "intellect" (manas), "ego-consciousness" (ahaṅkāra), and "judgement" (buddhi) stand as the doorkeepers. The latter three, taken together, constitute the so-called "inner organ" (antaḥkaraṇa);"

The experience of the senses are collected and registered through mind (*manas*), appropriated by ego-consciousness (ahaṅkāra) and then delivered to Judgement (*buddhi*). Since the "intellect" (manas) co-operates directly with the ten faculties, it is reckoned as number eleven and is termed "the inner sense" (antarindriya)[18]. In the monistic Vedānta school the inner organ is exactly the same as Mind (*manas*).

References

1. Chāndogya-unpaniṣad VII, 1, 6
2. ibid. VII, 26, 1
3. Bṛhadāraṇyaka-upaniṣad III 2, 13
4. Vaiśeṣika-sūtra I, 1, 5. The same theory is set forth in Praśastapāda's Padārthadharmasaṃgraha, chapter II. Cf. Radhakrishnan, Moore (eds): op. cit. p298
5. Vaiśeṣika-sūtra III, 2, 1. Detailed discussion follows (op. cit. III, 2ff)
6. Radhakrishnan, Moore op. cit. p400
7. Vyasa's Yogabhāṣya, IV, 16
8. Yogasūtra III, 19
9. Sāṃkhya-kārikā, 7
10. Soul (self), body, senses, objects of sense, intellect, mind activity, fault, transmigration [rebirth], fruit, pain, and release are the objects of right knowledge. (Nyāyasūtra, I, 1, 9)
11. Vaiśeṣika-sūtra VIII, 1, 2
12. ibid. VIII i, 4
13. Nyāya-sūtra I, 1, 16
14. Praśastapāda's opinions. I cite from Radhakrishnan, Moore op.cit. p408–409
15. ibid. p406–407
16. ibid. p409
17. Yogasūtra III, 11. Cf. Woods JH (1927) The Yoga System of Patañjali, vol 17. Harvard Oriental Series (Cambridge, Mass: Harvard University Press), p209
18. Zimmer H (1951) Philosophies of India. Pantheon Books, New York, p317f

IV. The Buddhist Approach to Dissolve Mind into Mental Functions

1. The Theory of "Non-Self"

Buddhist philosophy did not admit the existence of ego or soul as a substance. To make clear the teaching of Non-Ego, Buddhists set forth the theory at the Five Aggregates of Constituents (*skandhas*)[9] of our existence. Our human individual existence, the total of our mind and body, was divided into five groups of changing constituents. The five aggreates make up the individual. These Five Aggregates or Constituents are as follows:

(Provisional Reality = the Five Aggregates) (Fiction)

1. Corporeality or physical forms (*rūpa*)[2]
2. Feelings or sensations (*vedanā*)[3] (pleasant, unpleasant, neutral)
3. Ideations (sañña in Pāli, saṃjñā in Sanskrit)[4]
4. Latent formative forces or dispositions (saṅkhāra in Pāli, saṃskāra in Sanskrit)[5]
5. Consciousness (*viññāṇa* in Pāli, *vijñāna* in Sanskrit)[6]

In order to make this teaching slightly more tangible, we may cite the example of toothache. Normally, one simple says, "I have a toothache", but to the Buddhist thinkers this would have appeared as a very inconsistent way of speaking. Neither "I", nor "have", nor "toothache" are counted among the ultimate facts of existence (*dharmas*). In the Buddhist literature personal expressions are replaced by impersonal ones. Impersonally, in terms of ultimate events, this experience is divided up into the following five:

1. This here is the physical *form i.e.*, the tooth as matter:
2. There is a painful *feeling*:
3. There is a sight-, touch-, and pain perception (*ideation*) of the tooth; perception can exist only as ideation;
4. There is by way of volitional reactions; resentment at pain, desire for physical well-being, etc.
5. There is consciousness—an awareness of all the above-mentioned four[7].

The "I" of common sense talk has thus disappeared: It is not the ultimate reality. Not even its components are reality. One might reply, of course; an imagined "I" is a part of the actual experience. In that case, it would be placed in the category of consciousness, the last one of the five above-mentioned. But his consciousness is not ultimate reality. Our human experience is only a composite of the five aggregates (*skandhas*). None of the Five Aggregates is the self or soul (*attan* in Pāli, àtman in Sanskrit), nor can we locate the self or soul in any of them. A person is in process of continuous change, a flow, with no fixed underlying entity. In this way Buddhism swept away the traditional conception of a substance called "soul" or "ego", which had hithertofore dominated the minds of the superstitious and the intellectuals alike. Instead the teaching of *anattan*, non-self, has been held throughout Buddhism. These are of provisional reality. Once one attains enlightenment by true wisdom, they are brought to the state where they don't operate. So, they might be called "realms" (of functions), viewed from a higher point.

Buddhism believes that our existence is maintained and formed in the area of the five "components". Our existence is formed in the areas of these five different classes of provisional functions. The combination of everything that exists in such realms is provisionally called "self", "I" (ātman—the self or the ego) from the worldly, conventional point of view, but the subject of human existence cannot be included in any of the above-mentioned realms. Buddhism propounds with regard to material things (the first of the five):

1. Material things are impermanent:
2. What is impermanent is suffering:
3. What is suffering is not one's self: that is something other than one's self.
4. What is not one's self does not belong to one's self: in it one's self does not exist; it is not one's ātman.

The same form of argumentation is set forth regarding the other four constituents: feeling, ideations, dispositions, and consciousness. Everything that worldly people might consider the self or soul (ātman) is not the self (ātman) at all.

Thus, the Buddha explained[8] the non-perceptibility of the soul: "The physical form is not the eternal soul, for it is subject to destruction[9]. Neither feeling, nor ideation, nor consciousness, together constitute the eternal soul, for were it so, feeling, etc., would not likewise tend towards destruction".

In another passage it is taught: "Our physical form, feeling, perception, disposition and consciousness are all transitory, and therefore suffering is not permanent (and not good for our spiritual benefit). That which is transitory, suffering and liable to change is not the eternal soul. So it must be said of all physical forms whatever, past, present, or to be, subjective or objective, far or near, high or low: this is not mine, this I am not, this is not my eternal "soul"[10]. And the same assertion can also be said of feeling, ideation, dispositions, and consciousness also. All are impermanent; body, feeling, perception, dispositions, and consciousness, all these are suffering (dissatisfactory). They are all "non-self". Nothing of them is substantial. They are all appearances empty of substantiality or reality. There can be no individuality without a putting together of components. And this is always a process of "becoming": this is a becoming different: and there can be no becoming different without a dissolution, a passing away, or a decay, which sooner or later will inevitably come about.

Also, besides the theory of the Five Constituent Aggregates, early Buddhists set forth another theory of systems of "Realms" in which our cognitions and actions are formed. They are the sense of visual function, the sense of hearing, the sense of smell, the sense of taste, the sense of touch and Mind. They are called the Six Realms (or Situations). At the same time, corresponding to these six, another system of "Realms" was established. This is the system of the Realms of the Six Objects, which includes: visual forms, sounds, ōdour, taste, things to be touched, and things to be thought. In living human existence there is a continually succeeding series of mental and physical phenomena. It is the union of these phenomena that makes the individual. Every person, or thing, is therefore put together, a compound of components which change. In each individual, without exception, the relation of its components is always such, that no sooner has individuality begun than its dissolution, disintegration, also begins.

Many Buddhist terms are very difficult to translate into English. For certain technical terms there are no exact equivalents. The terms used in Western languages can only give a rough understanding of Buddhist teaching on this subject, but anyhow, the purport of this theory is rather simple. As we notice, in daily life, I assume that something is mine, or that I am something, or that something is myself. But this is wrong. Thus Buddhism denies the assumption of the existence of *ātman* (the self or the ego) as a metaphysical principle. Hence the thought of Buddhism is called the theory of "Non-Self". However, it never denies the self (ātman) itself. It merely insists that any object which can be seen in the objective world is no *ātman*. Regarding the question whether *ātman* exists or not, Buddhism gives no answer. The Buddha neither affirms nor denies the existence of *ātman*. He exhorts us to be philosophical enough to recognize the limits to philosophy. As body (corporality) is a name for a system of functions, even so *soul* is a name for the total of the states which constitute our mental existence. Without functions there can be admitted no soul. Therefore, it is not correct to understand Buddhism as the theory of non-existence of soul.

1. Or it might be translated as "constitutent aggregates". *Pañcaskandha* in Sanskrit, *go-un* or *go-on* in Japanese. "Khandha: A part of a whole thing, ingredients of the worldly existence; the constituents of the individual form, feeling, notion, mental dispositions, clear consciousness or discrimination." (S.C. Banerji: An Introduction to Pāli Literature, p. 141).

2. In many cases "physical form" (*rūpa*) (*shiki* in Japanese) means "Physical Form pertaining to the body" or "the body of a human being". Occasionally it is rendered as "matter", a rough translation. According to interpretations by later Abhidharma teachers, it can mean "matter", including everything both spiritual and material. Spiritual effects also were regarded as a kind of "latent matter".

3. *vedanā* (*ju* in Japanese). There are three kinds of feeling, *i.e.*, pleasant, unpleasant, and neutral. Later teachers of Abhidharma

and China interpreted it as "feeling, signifying the acceptance of impression withon one's consciousness".

4. *saṃjñā* (*sō* in Japanese), ideation, meaning: to form an image within one's consciousness, according to the *Ābhidharmika* and Chinese dogmaticians. The Pāli word *saññā* is occasionally translated as "conciousness". Consciousness is the concept closes to the concept of "self" or "soul". But early Buddhists made a distinction. To the question: "Is the consciousness (saññā) identical with a man's soul (*attan*), or is consciousness one thing, and the soul another?" The Buddha replied, "Granting a material soul, having form built up of the four elements, nourished by solid food; still some ideas, some stages of consciousness, would arise to the man, and others would pass away. On this account also, you can see how consciousness must be one thing, and soul another". (Rhys Davids TW (tr.) Dialogues of the Buddha. Part I, pp 252–253).

5. saṃkhāra (saṃskāra in Sanskrit, *gyō* in Japanese) is a different term to translate. "The confections" (Rhys Davids TW); "the predispositions" (Warren); "the constituent elements of character" (Rhys Davids TW). When we consider the interpretations by later Abidharma and Chinese teachers, we can translate it as "latent, formative, phenomena" or "formative forces", including activeness and latent formative forces". *Saṃkhāra* is explained by Warder AK (1963) Introduction to Pāli. Luzac p 277, as follows: "Force, energy, activity, combination, process, instinct, habit (a very difficult word to find an exact equivalent for; "force", with restricted technical sense attached to it, is probably the best. *Saṃskāra* means force or forces manifested in the combination of atoms into all the things in the universe, in the duration of such bominations—as in the life-span of a living being—and in the instincts and habits of living beings, which are to be allayed by the practice of meditation (*jhāna*). Samskāra is one of the five basic groups (*khandha*) of kinds of things in the universe: matter, sensation, perception and conciousness being the others.

6. vijñāna, (*shiki* in Japanese). Later Ābidharmika and Chinese teachers interpreted it as "cognition" denoting the act of distinguishing every object and recognizing it.

7. These illustrations were taken from E. Conze's Buddhims. cf. Saṃyutta-Nikāya, III, 46 etc.

8. In the Buddha's first sermon addressed to the five ascetics in Benares, (Dhammacakkappavattana-sutta), the non-perceptibility of the soul was set forth.

9. Or it can be translated as: "it is subject to destruction".

10. Vinaya, Mahāvagga I, 6, 38 f vol I, p 13 f Vinaya, Mahāvagga, I, 21. cf. Saṃyutta-Nikāya, VI, 54.

2. The Practical Implication of the "Non-Self" Theory

If we apply this method of analytical reflection properly, it must have a tremendous power to disintegrate unwholesome, selfish tendencies of human behaviour. Early and later Conversative (Hīnayāna) Buddhists thought seriously that the meditation on these component elements by itself alone can obviously uproot all the evil in our hearts. It was believed that it was bound to contribute to our spiritual development to the extent that, when it is repeated often enough, it will set up the habit of viewing all things impersonally, which means apart from our selfish tendency. This way of thinking may be applied with efficiency by modern psychiatrists.

3. Analysis of Mind by Way of Buddhist Psychology

(i) The Analysis by the Sarvāstivādins — The Pluralistic Concept of Mind

The psychological analysis of one's own existence was not yet systematized in Early Buddhist scriptures. It was owing to the efforts by teachers of dogmatics (Abhidharma teachers) that various aspects of human existence were analyzed and schematized very elaborately. The school which was most influential and powerful among the school of Conservative Buddhism (Hīnayāna) was the Sarvāstivādins (lit. "the school which asserts that all (dharmas) exist"). Dharmas are constituents of our existence. This school admitted these Five Aggregates (*Skandhas*), which were also called "dharmas". The Sarvāstivādins advocated that, although things in the phenomenal world may vanish, these *dharmas* or entities exist actually. (In this respect, compare Husserl's phaenomenology).

It was expecially in the Sarvāstivāda school that developed an elaborate system of psychological analysis, according to which all constituent elements of human existence were classified as 75. These 75 are divided in two major groups: 1. saṃskṛta: Cooperating, impermanent elements, and 2. asaṃskṛta: non-cooperating, immutable elements. The former is divided in four major groups.

A. Materral Elements (*rūpa*)

1. cakṣur-indriya: visual organ
2. śrotra-indriya: auditory organ
3. ghrāṇa-indriya: olfactory organ
4. jihvā-indriya: taste organ
5. kāya-indriya: tactile organ
6. rūpa-viṣaya: visual sense-data
7. śabda-viṣaya: auditory sense-data
8. gandha-viṣaya: olfactory sense-data
9. rasa-viṣaya: taste sense-data
10. spraṣṭavya-viṣaya: tactile sense-data
11. avijñapti-rūpa: unmanifested matter which is the vehicle of moral qualities

B. Mind (*citta*)

This is pure consciousness without content.

C. The forty-six mental elements (*caitta-dharma*) or faculties intimately combining with the element of consciousness (citta-saṃprayukata-saṃskāra).

These are divided into six groups as follows:

1. Ten "General Functions", *i.e.* general mental faculties present in every moment of consciousness (citta-mahābhūmikāḥ [dharmāḥ])

a. vedanā: faculty of feeling (pleasant, unpleasant, and indifferent)[1]

b. saṃjñā: faculty of ideation[2]

c. cetanā: faculty of will, causing action of mind[9]

d. sparśa: sensation, caused by "contact" between object, sense-organ, and consciousness[4]

e. chanda: faculty of desire[5]

f. prajñā (or mati): faculty of intelligence (discriminative knowledge of *dharmas*)[6]

g. smṛti: faculty of conscious memory[7]

h. manasikāra: faculty of attention[8]

i. adhimokṣa: faculty of ascertainment (or dicisive knowledge)[9]

j. samādhi: faculty of concentration[10]

2. Ten "General Good Functions", *i.e.*, universally "good" moral faculties, present in every good moment of consciousness (kuśalamahābhūmikāḥ) [dharmāḥ]

a. śraddhā: faculty of belief, causing mind to be pure and joyful[11]

b. vīrya: faculty of courage in good actions[12]

c. upekṣā: faculty of equanimity of indifference[13]

d. hrī: faculty of modesty, being respectful to virtuous persons[14]. (According to some teachers:) faculty of modesty, being ashamed with reference to oneself[15]

e. apatrāpya (or apatrapā): faculty of awefulness with regard to sins[16]. (According to some teachers:) faculty of feeling disgust with reference to other peoples objectionable actions[17]

f. alobha: faculty of non-greediness[18]

g. adveṣa: faculty of non-malevolence[19]

h. avihiṃsā: faculty of causing no injury[20]

i. praśrabdhi: faculty of mental dexterity or mental suitability for any action[21]

j. apramāda: faculty of making endeavour to acquire good virtues[22]

3. Six "General Functions of Defilement", *i.e.*, universally "defiled" elements present in every unfavourable moment of consciousness (kleśamahābhūmikāḥ [dharmāḥ]).

a. moha (or avidyā): faculty of infatuation or ignorance[22]

b. pramāda: faculty of laziness (*i.e.* no practice of good virtues)[23]

c. kauśīdya: faculty of mental indolence[24]

d. āśraddya: faculty of non-believing[25]

e. styāna: faculty of sloth or indolence, inactive temperament[26]

f. auddhatya: faculty of being agitated and disturbed of mind[27]

These six faculties are not always absolutely bad; they may be sometimes indifferent for the spiritual progress, but nevertheless they always function with a selfish tendency.

4. Two "General Functions of Evil", *i.e.* universally "bad" elements present in every bad moment of consciousness (akusála-mahābhūmikau [dharmau])[28]

a. āhrīkya: faculty of irreverence, lack of modesty. (According to the orthodox teaching:) not being respectful to virtuous persons. (According to some teachers:) not being ashamed with reference to onseself[29]

b. anapatrāpya (or anapatrapā): faculty of not feeling aweful with regard to sins. (According to some teachers:) faculty of not feeling disgust at offences done by other[30]

5. Ten "Minor Functions of Defilement", *i.e.*, vicious elements of limited occurrence (parītta-bhumikā[31] upakleśāḥ[39]). They occur occasionally.

a. krodha: faculty of anger[32]

b. mrakṣa: faculty of hypocrisy (concealing one's own sins)[33]

c. mātsarya: faculty of stinginess[34]

d. īrṣyā: faculty of jealousy[35]

e. pradāśa: faculty of insisting on objectionable things[36]

f. vihiṃsā: faculty of causing injury[37]

g. upanāha: faculty of resentment[38]

h. māyā: faculty of deceit[39]

i. sāthya: faculty of fraudulance[40]

j. mada: faculty of complacency[41]

6. Eight "Indeterminate Functions", *i.e.*, elements not having any definite place in the above system, but capable of entering into various combinations (aniyatāḥ)[43]

a. kaukṛtya: faculty of repenting[44]

b. middha: faculty of drowsiness[45]

c. vitarka: faculty of reflection[46]

d. vicāra: faculty of subtle investigation[47]

e. rāga: faculty of attachment by mind[48]

f. pratigha: faculty of hatred[49]

g. māna: faculty of arrogance[50]

h. vicikitsā: faculty of doubting[51]

D. Forces which can neither be included among material nor among spiritual elements (citta-viprayuktāḥ saṃskārāḥ)[52]

1. prāpti: "acquisition", a force which effects the acquisition of the elements in an individual existence[53]

2. aprāpti: "non-acquistition", a force which occasionally keeps some elements in abayance in an individual existence[54]

3. nikāya-sabhāgatā: "similarity of existence", a force producing generality or homogeneity of existences[55]

4. āsaṃjñika: a force which transfers an individual into the realm of the unconscious trance[56]

5. asaṃjñi-samāpatti: a force stopping consciousness and producing the annihilation trance[57]

6. nirodha-samāpatti: a force stopping consciousness and producing the annihilation trance (the highest trance)[58]

7. jīvita: the force of life-duration[59]

8. jāti: the force of origination[60]

9. sthiti: the force of subsistence[61]

10. jarā: the force of decay[62]

11. anityatā: the force of extinction[63]

12. nāma-kāya: the force imparting significance to words[64]

13. pada-kāya: the force imparting significance to sentences[65]

14. vyañjana-kāya: the force imparting significance to articulate sounds[66]

E. Immutable elements (asaṃskṛta-dharma)

1. ākāśa: space for all dharmas[67]

2. pratisaṃkhyā-nirodha: the extinction of the manifestations of elements through the action of discriminative knowledge[68]

3. apratisaṃkhyā-nirodha: the extinction of the manifestations of elements due to lack of productive causes, not through the action of discriminative knowledge[69]

In the philosophical system of this school, all the *dharmas* are classified in the 75 categories, as are explained above. They do not depend on each other, maintaining their respective, independent existence. Each *dharma* comes to appear, and then vanishes in our consciousness, but a *dharma* itself preserves its own self-identity throughout the past, the present and the future. This theory was called the theory of "the permanent existence of the essence (entity) of each *dharma*"[70] or "the theory of the existence of a dharma as a substance throughout the three divisions of time, *i.e.* the past, the present and the future"[71].

The common features which the Sarvāstivādin theory shares with the Platonic theory of ideas was already pointed out by such Russian scholars as Otto Rosenberg and Th. Stcherbatsky[72].

This school was founded by a scholar named Kātyāyanīputra (2nd century B.C.) who wrote the Abhidharmajñānaprasthāna-śāstra, the fundamental text of this school. It maintained the theory that all dharmas, which constitute a human existence, such as the five Aggregates (skandhas), the twelve Regions (āyatanas), the eighteen Elements (dhātus) etc., *i.e.*, systems of dharmas in their respective viewpoints, do really exist. In this school the term "dharma" meant something like an essence (Wesen, in German). According to this school, these dharmas exist as substances (dravyataḥ sat)[2], or exist essentially (svalakṣaṇataḥ sat)[3].

References

1. vedanā trividho 'nubhavaḥ, sukho duḥkho 'nubhayaś ca. (Abhidharmakośabhāṣya ed. by Pradhan, p 54, 11. 19–20. Hsüan-tsang's Chinese translation, vol 4, Taishō Tripiṭaka, vol 29, p 19 a)
2. "saṃjñā" saṃjñānam viṣayanimittodgrahaḥ (AKBh p 54, 11. 20–21) (Chinese tr. p 19 a)
3. "cetanā" cittābhisaṃskāro manaskarma (AKBh, p 54, 1. 20. Chinese tr. p 19 a)
4. "sparśa" indriya-viṣaya-vijañāna-sannipātajā (AKBh p 54, 1. 21. Chinese tr. p 19 a)
5. "chandaḥ" kartṛkāmatā (AKBh p 54, 1. 21. Chinese tr. 19 a)
6. "matiḥ" prajñā dharma-pravicayaḥ (AKBh p 54, 1. 22. Chinese tr. p 19 a)
7. "smṛtir" ālambana-asampramoṣaḥ (AKBh p 54, 1. 22. Chinese tr. p 19 a)
8. "manaskāraś" cetasa ābhogaḥ (AKBh p 54, 1. 22. Chinese tr. p 19 a)
9. "adhimokṣo" 'dhimuktiḥ (AKBh p 54, 1. 23. Chinese tr. p 19 a)
10. "samādhiś" cittasyaikāgratā (AKBh p 54, 1. 23. Chinese tr. p 19 a)
11. "śraddhā" cetasaḥ prasādaḥ. Another theory runs as follows: satya (*i.e.* catvāri satyāni)-ratna (*i.e.* Buddha, Dharama, Saṃgha)-karma-phala-adhisaṃpratyaya ity apare (AKBh p 55, 11. 6–7. Chinese tr. p 19 b)
12. "vīryam" cetaso 'bhyutsāhaḥ (AKBh p 55, 1. 23. Chinese tr. p 19 b)
13. "upekṣā" cittasamatā citta-anābhogatā (AKBh p 55, 1. 16. Chinese tr. p 19 b)
14. "ahrīr" agurutā (Abhidharmakośa, II, v. 32). guṇeṣu guṇavatsu cāgauravatā apratīśatā abhayam avaśavartitā āhrīkyaṃ gaurava-pratidvaṃdvo dharmaḥ. (AKBh p 59, 11. 17–18). viparyayeṇa hrīr apatrāpyam ca veditavyam (AKBh p 60, 1. 3). gurutvaṃ hrīḥ. gauravaṃ hi nāma sapratīśatā. tatpūrvikā ca lajjā hrīḥ. ato na gauravam eva hrīr ity apare (ibid. p 60, 11. 14–15)
15. anye punar āhuḥ—ātmāpekṣayā doṣair lajjanam "āhrīkyam"... iti. (ibid. p 24) "hiri (fem.)—modesty, self-respect, conscience"
16. avadye bhayādarśitvam atrapā (Abhidhrmakośa, II, v. 32). avadyaṃ nāma yadgarhitaṃ sadbhiḥ. tatrābhayadarśitā 'napatrāpyam. "bhayam" atrāniṣyaṃ phalaṃ bhīyate 'smād iti. (AKBh p 59, 11. 20–21) viparyayeṇa hrīr apatrāpyaṃ ca veditavyam (ibid. p 60, 1. 3.)
17. anye punar āhuḥ—doṣair alajjanam ... parāpekṣayā "napatrāpyam" (AKBh p 59, 1. 24)
18. AKBh p 55, 1. 22
19. AKBh p 55, 1. 22
20. "avihiṃsā" aviheṭhanā (AKBh p 55, 1. 23. Chinese tr. p 19 b)
21. "praśrabdhiś" citta-karmaṇyatā (AKBh p 55, 1. 8. f)
22. "apramādaḥ" kuśalānāṃ dharmāṇāṃ bhāvanā. Another interpretation by another school (outside the Sarvāstivādins) is: cetasa ārakṣaḥ (AKBh p 55, 1. 7 f)
23. "pramādaḥ" kuśalānāṃ dharmānām abhāvanā 'pramāda-vipakṣo dharmaḥ (AKBh p 56, 11. 6–7. Chinese tr. p 19 c)
24. "kauśīdyaṃ" cetaso nābhyutsāho vīryavipakṣaḥ (AKBh p 56, 1. 7, Chinese tr. p 19 c)
25. "āśraddhyam" cetaso 'prasādaḥ śraddhā-vipakṣaḥ (AKBh p 56, 11. 4–7 Chinese tr. p 19 c)
26. "styānam" katamat? yā kāyagurutā cittagurutā kāya-akarmaṇyatā citta-akarmaṇyatā, kāyikaṃ styānaṃ caitasikaṃ styānam iti ... (AKBh p 56, 1. 8 f. Chinese tr. p 19 c)
27. "auddhatyam" ... cetaso 'vyupaśamaḥ (AKBh p 56. 1. 10. Chinese tr. p 19 c)
28. Abhidharmakośa II, 26
29. See n. 14 and 15
30. See n. 17 and 18
31. paritta-kleśa-bhūmikāḥ, Abhidharmakośa II, 27
32. unpakleśaḥ, AKBh, p 58, 1. 12
33. vyāpāda-vihiṃsā-varjitaḥ sattvāsattvayor āghātaḥ "krodhaḥ" (AKBh p 312, 1. 18. Chinese tr. p 109 b)
33′. avadya-pracchādanaṃ "mrakṣaḥ" (AKBh p 312, 1. 18. Chinese tr. p 109 b)
34. dharmāmiṣakauśala-pradāna-virodhī cittāgraho "mātsaryam" (AKBh p 312, 11. 15–16. Chinese tr. p 109 b) Cf. Adhidharmakośa V, 47
35. para-saṃpattau cetaso vyāroṣa "īrṣyā". (AKBh p 312, 1. 15. Chinese tr. 109 b). Cf. Abhidharmakośa, V, 47)
36. sāvadyavastudṛḍhagrāhitā "pradāśo" yena nyāyasaṃjñaptiṃ na gṛhṇāti (AKBh p 313, 11. 14–15. Chinese tr. p 109 c)
37. viheṭhanaṃ "vihiṃsā" yena prahārapāruṣyādibhiḥ parān viheṭhayate (AKBh p 313, 11. 15–16. Chinese tr. p 109 c)
38. āghātavastu-bahulīkāra "uppanāhaḥ" (AKBh p 313, 1. 15. Chinese tr. p 109 c). Cf. Abhidharmakośa V, 49 and 50
39. māyā (Abhidharmakośa V, 50). In the Sanskrit text of the Abhidharmakośabhāṣya there is no explanation on the term "māyā", but in the Chinese translation by Hsüan-tsang (p 109 c) it is said: "māyā means 'to deceive others'"
40. citta-kauṭilyaṃ "śāṭhyaṃ" yena yathābhūtaṃ nāviṣkaroti vikṣipaty aparisphuṭaṃ vā pratipadyate (AKBh p 313, 11. 13–14. Chinese tr. p 109 c). Cf. Abhidharmakośa V, 51
41. "Madas" tv svadharmeṣv eva raktasya yac cetasaḥ paryādānam, yathā madyaja evaṃ rāgajaḥ. saṃpraharṣaṇa-viśeṣo "mada" ity spare (AKBh p 60, 11. 16–17. Chinese tr. p 21 c). Cf. Abhidharmakośa V, 50
42. aniyatāḥ, AKBh ad II, 27; p 57, 1.8; AKV p 132
43. It is likely that various sorts of defilements were common to Buddhism and Jainism. rāgo doso moho annāṇam jassa avagayaṃ hoti, āṇāe roento so khalu ānāruī nāmaṃ. (Uttarajjhāyā 28, 20). "He who has got rid of love, hate, delusion and ignorance, and believes because he is told to do so believes by command." This verse stresses getting rid of the kaṣāyas and of ingorance
44. kim idaṃ kaukṛtyaṃ nāma? kukṛtasya bhāvaḥ "kaukṛtyam. iha tu punaḥ kaukṛtyālambano dharmaḥ "kaukṛtyam" ucyate cetaso vipratisāraḥ (AKBh p 57, 11. 18–19. Chinese tr. p 20 b) Cf. Abhidharmakośa II, 28

45. Kāyasaṃdhāraṇa-asamarthaś cittābhisaṃkṣepo "middham" (AKBh p312, l. 17. Chinese tr. p109b). Cf. Abhidharmakośa V, 47
46. cittaudārikatā vitarkaḥ. cittasūkṣmatā vicāraḥ (AKBh p60, l. 22. Chinese tr. p21b) Cf. Abhidharmakośa II, 33
47. Cf. n.46
48. sukhāyāṃ vedeanāyāṃ rāgo 'nuśete (AKBh p39, ll. 19–20. Chinese tr. p14a)
49. anuśete duḥkhāyām (vedanāyām) pratighaḥ (AKBh p39, l. 20. Chinese tr. p14a) Cf. AKBh p277, ll. 11; 16; p279, l. 17
50. yena kenacit parati viśeṣaparikalpena cetasa unnatiḥ "mānaḥ" (AKBh p60, l. 16. Chinese tr. p21c). Cf. Abhidharmakośa II, 33
51. vicikitsa, cf. p277, l. 17; p279, l. 17; p307; l. 16etc
52. or viprayuktāḥ saṃskārāḥ (Abhidharmakośa II, 35). ime saṃskārā na cittena saṃprayuktā naca rūpasvabhāvā iti "cittaviprayuktā" ucyate (AKBh p62, l. 14). Chinese tr. p22a)
53. There are two kinds of prāpti: (1) lābha and (2) samanvaya (Abhidharmakośa II, 36). dvividhā hi "prāptir" aprāptavihīnasya ca lābhaḥ pratilabdhena ca samanvāgamaḥ. viparyayād "aprāptir" iti siddham. kasya punar ime prāpty-aprāptī? prāpty-aprāptī svasaṃtānapatitānāṃ, na para-saṃtāna-patitānām, ... nirodhayoḥ (of apratisaṃkhyānirodha and pratisaṃkhyānirodha). (AKBh p62, ll. 16–23. Chinese version, p22a)
54. Cf. n.53
55. "sabhāgatā" sattvasāmyaṃ (Abhidharmakośa II, 419). sabhāgatā nāma dravyam. sattvānāṃ sādṛśyaṃ nikāyasabhāga ity asyāḥ śāstre saṃjñā. (AKBh p67, ll. 12–13. Chinese tr. p24a)
56. asaṃjñisattveṣu deṣūpapannānāṃ yaś cittacaittānām nirodhas tad "āsaṃjñikaṃ" nāma dravyaṃ yena cittacaittā anāgate 'dhvani kālāntaram saṃnirudhyante notpattuṃ labhante (AKBh p68, ll. 13–14. Chinese tr. p24b). Cf. Abhidharmakośa II, 41
57. asaṃjñināṃ samāpattir, asaṃjñā vā (samāpattir) ity asaṃjñisamāpattiḥ. sā'pi cittacaittānāṃ nirodhaḥ ... sā tu samāpattir antye dhyāne (= caturthe dhyāne) (AKBh p69, l. 2–5. Chinese tr. 24c)
58. nirodhasamāpattir ... nirodhaś cittacaittānāṃ ... vihārārthaṃ śāntavihāra-saṃjñā-pūrvakeṇa manasikāreṇa enāṃ samāpadyante (AKBh p69, l. 24, p70, l. 5. Chinese tr. p24c–25a)
59. āyur "jīvitam" ādhāra uṣmavijñānayor ... (Abhidharmakośa II, 45. p73. Chinese tr. 26a)
60. "jātis" taṃ dharmaṃ janayati (AKBh p75, l. 19. Chinese tr. p27a)
61. "sthitiḥ" taṃ dharmaṃ sthāpayati (AKBh p75, l. 19. Chinese tr. p27a)
62. "jarā" taṃ dharmaṃ jarayati (AKBh p75, l. 19. Chinese tr. p27a)
63. "anityatā" taṃ dharmaṃ vināśayati (AKBh p75, l. 19–20. Chinese tr. p27a)
64. saṃjñā-karaṇaṃ "nāma". tadyathā rūpaṃ śabda ity evamādiḥ ... eṣāṃ ca saṃjñādīnāṃ samuktayo nāmādikāyāḥ. uca samavāye paṭhati. tasya samuktir ity etad rūpaṃ bhavati. yo 'rthaḥ samavāya iti so 'rthaḥ samuktir iti. tatra nāmakāyas tadyathā rūpaśabdagandharasa spraṣṭavyānīty evamādi (AKBh p80, ll. 13–20. Chinese tr. p29a)
65. vākyaṃ "padaṃ" yāvatā 'rthaparisamāptis tadyathā "anityā bata saṃskārā" ity evamādi yena kriyāguṇakālasaṃbandhaviśeṣā gamyante. padakāyaḥ tadyathē "sarvasaṃskārā anityāḥ sarvadharmā anātmānaḥ śāntaṃ nirvāṇam" ity evamādi. (AKBh p80, ll. 14–15. Chinese tr. p29a)
66. "vyañjanam" akṣaraṃ tadyathā a ā ityevamādi ... "vyañjanakāyas" tadyathā ka kha ga gha ṅa-ity evamādi. (AKBh p80, ll. 21–22. Chinese tr. p29a)
67. "ākāśam" anāvṛtiḥ. (Abhidharmakośa I, 5) anāvaraṇa-svabhāvaṃ ākāśaṃ yatra rūpasya gatiḥ (AKBh p3, l. 23. Chinese tr. p1c)
68. "pratisaṃkhyā-nirodho" yo visaṃyogaḥ (Abhidharmakośa I, 6). yaḥ sāsravair dharmair visaṃyogaḥ sa pratisaṃkhyā-nirodhaḥ. duḥkhādīnām āryasatyānāṃ pratisṃkhyānaṃ "pratisaṃkhyā" prajñā-viśeṣas, tena prāpyo nirodhaḥ "pratisaṃkhyā-niriodhaḥ" (AKBh p4, ll. 1–2. Chinese tr. p1c)
69. utpāda-atyanta-vighno 'nyo nirodho 'pratisaṃkhyayā (Adhidharmakośa I, 6). anāgatānāṃ dharmāṇām utpādasyātyanta-vighnabhūto visaṃyogād yo 'nyo nirodhaḥ so 'pratisaṃkhy ā-nirodhaḥ nāsau pratisaṃkhyayā labhyate, kiṃ tarhi 'pratyaya-vaikalyāt (AKBH p4, ll. 11–12. Chinese tr. p1c)
70. svabhāvah sarvadā cāsti (Abhidharmakośa-vyākhyā, op. cit. p472, l. 25)
71. traiyadhvika (= sarve saṃskṛtā dharmāḥ)
72. Rosenberg O (1924) Problem der Buddhistischen Philosophie, 1918. Translated into German, Heidelberg. Stcherbatsky Th op. cit

V. The Analysis by Buddhist Idealism (the Vijñānavāda)

The above-mentioned scheme of 75 elements was developed and enlarged to that of 100 elements by the Yogācāra idealists. Their standpoint is called Vijñānavāda (*i.e.* Consciousness-only Theory).

One item which is noteworthy from the psychological viewpoint is that the Vijñānavāda advocated the theory of the Eightfold Consciousness: 1st–5th; the First Five Consciousnesses: visual consciousness, auditory consciousness, our consciousness, taste consciousness, touch consciousness. 6th, the conscious mind. 7th, the subconscious mind (the substrate of self-consciousness). 8th, the Store-Consciousness (ālaya-vijñāna), the fundamental consciousness.

According to the orthodox thought of Dharmapāla (530–560 A.D.) which was conveyed at Hsuan-tsan to China and finally to Japan, as the Fa-hsiang (Hossō) school, these are separate consciousnesses as different entities and the first seven of these consciousnesses are collectively termed "the transformed consciousnesses" (pravṛttivijñāna, tenjiki in Japanese). The Shē-lun school of China regards the Store-Consciousness that has become pure and taintless as Thusness (tathatā) and gave it a special name, "Taintless Consciousness" (amala-vijāna), which was designated as the ninth consciousness.

The last one is so to speak the fundamental Original mind.

Generally speaking, Buddhist psychology was highly tinged with ethical and soteriological evaluation. Analytical studies on mind have been carried on in temples of Nara, the ancient capital of Japan, such as Hōryūji, Yakushiji, Kōfukuji and so on.

VI. Mind as the Metaphysical Fundamental Principle

In order to exhort disciples to practize meditations, a sort of idealism that all the universe is nothing but the outcome of Mind (cittamātraka) was strenuously taught[1] in the Daśabhūmika-sūtra of India and the Hua-yen school of China based on this scripture. This tendency gave rise to the idea of the Original Pure Mind in some schools of Mahayana Buddhism. According to the Awakening of Faith[2] ascribed to Aśvaghoṣa, the ultimate truth is the Mind of All Living Beings which comprises all worldly and non-worldly things.

The concept of the Originally Pure Mind, first held in Mahayana Buddhism, was introduced into Chinese philosophy, and especially into Neo-Confucian thought, and so the school of Mind came into existence. The pioneer of this trend was Ch'eng Hao (程颢, 1033-1085)[3]. The strong ethical tone of Ch'eng Hao's thought and his vigorous reaffirmation of life and the natural order mark him as truly within the Confucian tradition. However, the influence of Taoism and Buddhism, to which Ch'eng Hao had given years of study, is clearly discernible in his subjectivism and his idea of mental composure. He advocated "the way to preserve humanity". "As humanity is preserved, the self and the other are then identified. For our innate knowledge of good and innate ability to do good are part of our original nature and cannot be lost. However, because we have not got rid of the mind dominated by old habits, we must preserve and exercise our original mind, and in time old habits will be overcome."

Buddhist idealism gave rise to the idealistic philosophy of Neo-Confucianism in China. According to the philosophy of Chu Hsi (1130–1200)[4], there is an immaterial and immutable principle inhering in all things, which gives them their form and constitutes their essence. This principle in man is his true nature, fundamentally good. Man's mind, moreover, is in essence one with the mind of the universe, capable of entering into all things and understanding their principles. Chu Hsi believed in the perfectability of man, in the overcoming of those limitations or weaknesses which arise from an imbalance in his physical endowment. His method was the "investigation of things" that is, the study of their principles, and also self-cultivation to bring one's conduct into conformity with the principles which should govern it. Eventually, Chu asserted, persistent effort in this direction would result in everything's becoming suddenly clear and the full enlightenment of the sage being attained.

When Chu Hsi advocated investigating the principle in individual things, it meant that the principle in each individual thing is to be sought with the mind, thus separating the mind and principle into two. Against this theory Wang Yang-ming (1472–1529) advocated the theory of the Identification of Mind and Principle. He says,

"The innate knowledge of mind is the same as the principle of Heaven. When the principle of Heaven in the innate knowledge of my mind is extended to all things, all things will attain their principle. To extend the innat knowledge of my mind means extension of knowledge, and all things attaining their principle means investigation of things. In these the mind and principle are combined as one."

Wang Yang-ming went so far as to say[5],

"there is nothing under heaven that is external to the mind."

"Man is the mind of Heaven and Earth. And what is it in man that is called his mind? It is simply the spirituality or consciousness. In Heaven and Earth there is one spirituality or consciousness. But because of his bodily form, man has separated himself from the whole. My spirituality or consciousness is the ruler of Heaven and Earth, spritis and things. If Heaven, Earth, spirits, and things are separated from my spirituality or consciousness, the cease to be. And if my spirituality or consciousness is separated from them, it ceases to be also. Thus they are all actually one body, so how can they be separated?"

The universe is a spiritual whole, in which there is only one world, the concrete actual world that we ourselves experience. Wang Shou-jen also maintains that mind is Li: "Mind is Li (Reason). How can there be affairs and Li outside the mind?" (Record of Instructions, pt. 1). Again "The substance of the mind is the nature and the nature is Li." According to Wang Yang-min's system, if there is no mind, there will be no Li. Thus the mind is the legislator of the universe and is that by which the Li are legislated.

The thought to regard mind as the fundamental principle appeared in Japan also. Mind was eulogized by Eisai, (1141–1215), the twelfth century Zen master of Japan, as the fundamental principle of all the universe.

"Great is Mind. Heaven's height is immeasurable, but Mind goes beyond heaven; the earth's depth is also unfathomable, but Mind reaches below the earth. The light of the sun and moon cannot be outdistanced, yet

Mind passes beyond the light of sun and moon. The macrocosm is limitless, yet Mind travels outside macrocosm. How great is Space! How great the Primal Energy! Still Mind encompasses Space and generates the Primal energy. Because of it heaven covers and earth upbears. Because of it the sun and moon move on, the four seasons pass in succession, and all things are generated. Great indeed is Mind! Of necessity we give such a name to it, yet there are many others: the Highest Vehicle, the First Principle, the Truth of Inner Wisdom, the One Reality, the Peerless Bodhi, the Way to Enlightenment and Insight of Nirvāna."

According to him Śākyamuni Buddha transmitted his truth of the Mind to later masters, calling it a special transmission not contained in the scriptures[6].

It means that Truth as is called Mind is ineffable; it cannot be grasped with words. Only by means of Zen meditation it can be apprehended. Nakae Tōju (1608–1648), the Japanese Confucianist, admitted the Divine Light in the mind of a human being[7].

"The superior man will be watchful over those inmost thoughts known to himself alone. In his everyday thinking, he will not think anything for which he would have to fear if brought into the presence of the Divine. In his everyday actions he will not perform an act of which he might be ashamed if it were known to others. By mistake an evil idea may arise, a wrong deed may present itself; but since there is within the mind a divine awareness illuminating it, what we call "enlightenment" will come. Once this realization occurs, rectification will follow, the evil idea and wrong deed will disappear, and the mind will revert to its normal state of purity and divine enlightenment. The ordinary man, unfortunately, continues to think such evil thoughts and goes on doing what he knows is wrong. Nevertheless, since the divine light in the mind makes the man aware that he is doing wrong, he tries to hide it. In everybody's mind there is this divine light, which is one with the Divinity of Heaven, and before which one stands as if in a mirror, with noting hidden either good or bad".

According to Nakae, the absolute is the mind of the Supreme Lord, and the mind of a human being is an outcome of the Divine Mind.

"The Supreme Lord Above is infinite and yet He is the final end of all. He is absolute truth and absolute spirit. All forms of ether are His form; infinite principle is His mind. He is greater than all else and yet there is nothing smaller. That principle and that ether are self-sustaining and unceasing. Through their union He produces lives throughout all time, without beginning or end. He is the father and mother of all things. Through division of His form He gives form to all things; through division of His mind He gives all things their nature. When form is divided, differences result; when mind is divided; the minds remain the same[8]".

Some Buddhist metaphysicians asserted that mind as the fundamental principle is empty, not a substance. Emptiness itself is the absolute. This theory was adopted by some Buddhist Idealists (nirākāravijñānavādins), and influenced Chinese Buddhist philosophers greatly. Fa-wen (法温), the fourth-century Buddhist philosopher of China, advocated the theory of No Mind or the Emptiness of Mind. This idea was very influential in later days. The "mindlessness" which transcends mind, the moments of "no-action" which excite greater interest than those of action, the mind which controls all the powers—all these are familiar ideas of Zen. These ideas greatly influenced various aspects of Japanese arts. To illustrate, one can trace it in the Noh plays as can be seen in writings of Zeami, The Noh master[9].

"Sometimes spectators of the Nō say, "The moments of 'no-action' are the most enjoyable." This is an art which the actor keeps secret. Dancing and singing, movements and the different types of miming are all acts performed by the body. Moments of "no-action" occur in between. When we examine why such moments without actions are enjoyable, we find that it is due to the underlying spiritual strength of the actor which unremittingly holds the attention. He does not relax the tension when the dancing or singing come to an end or at intervals between the dialogue and the different types of miming, but maintains an unwavering inner strength. This feeling of inner strength will faintly reveal itself and bring enjoyment. However, it is undesirable for the actor to permit this inner strength to become obvious to the audience. If it is obvious, it becomes an act, and is no longer "no-action". The actions before and after an interval of "no-action" must be linked by entering the state of mindlessness in which one conceals even from oneself one's intent. This, then, is the faculty of moving audiences, by linking all the artistic powers with one mind."

The ultimate state of mind in Noh plays is "mindlessness", and yet the most important is mind.

"In the art of the Nō too, the different sorts of miming are artificial things. What holds the parts together is the mind. This mind must not be disclosed to the audience. If it is seen, it is just as if a marionette's strings were visible. The mind must be made the strings which hold together all the powers of the arts. If this is done the actor's talent will endure."

References

1. In the chapter of the sixth bhumi of the Daśabhūmika-sūtra. Kumataro Kawada (in Engl.) in Journal of Indian and Buddhist Studies (1962) vol 10, No 1, Jan., p 329 f
2. Mahāyāna-śraddhotpāda-śāstra (大乘起信論). Taisho Tripiṭaka, vol XXXII, p 575 f
3. De Bary: Sources of Chinese Tradition, op. cit. pp 559–560
4. ibid. pp 577–578
5. Fung Yu-lan (1958) A Short History of Chinese Philosophy. MacMillan, New York, pp 309–310
6. Taishō Daizōkyō, vol 80, p 2. Tsunoda op. cit. p 242 f
7. Tetsujiro Inoue: Yomei Gakuha no Tetsugaku, pp 81–85. Tsunoda op.cit. p 382
8. Tsunoda op. cit. p 382
9. ibid. p 291 f

VII. Conclusion

We cannot represent Eastern theories on mind in a brief sentence. Eastern thought has displayed quite a variety on the problem of mind. I have just made brief, introductory remarks on this problems, which I hope may help scholars launch further investigations in their own ways.

Correspondence: Prof. Dr. H. Nakamura, 4-37-15 Kugayama, Suginami-Ku, Tokyo 168, Japan.

Acta Neurochirurgica, Suppl. 44, 33–38 (1988)

The Dynamics of Personality. An East European View

J. Kelemen

Department of Philosophy, Faculty of Arts, Eőtvős Loránd University, Budapest, Hungary

Summary

As the debates on fundamental issues of the human sciences have shown, actually two significant pictures of man are opposed to each other: the scientific-biological and the historicist conceptions. The second one is supported, for instance, by the Budapest School of Philosophy (with Lukács and his followers) and the Soviet School of Psychology (Vigotsky, Luria, Leontiev and their pupils). Summing up the doctrines of these two currents of thought the author argues in favour of the historicist conception and indicates, what is implied, in different fields of application (mind-body problem, linguistic theories, neuropsychic diseases, mental disorders), by the acceptance of this point of view. Moral issues, concerning psychiatric treatment, are discussed, too, and a Kantian solution of them is defended.

Keywords: Budapest School of Philosophy; Soviet School of Psychology; conceptions (pictures) of man; mind-body problem; social character of language; moral problems of psychiatry; humanist psychiatry.

(0.0) Is it possible to frame different concepts of personality contingent upon geographical, political, and cultural differences? Is there an European or an Asian, a Western or an Eastern conception of personality? Is a regional approach of the theory of personality compatible with the claim to universality of sciences, especially those concerning the biological aspects of man?

A simple answer in the affirmative would sound paradoxical, but the questions are not unintelligible. They show that the picture we draw of the structure and dynamics of personality cannot be supported only by empirical considerations. The underdeterminateness of scientific theories, including even physical ones, obtains to a greater degree in psychology, psychiatry, and generally, in all sciences dealing with conscious human behaviour. There is more than one theory compatible with a given set of observed data; the choice of a theory, therefore, is determined, over and above empirical verification, by our values, traditions, philosophical and ethical considerations.

Our choice among theories of personality depends, apart from scientific arguments, on an underlying, general picture of man. The debates on fundamental issues of the human sciences, which brought about a theoretical and institutional crisis in psychiatry, have shown that actually two significant pictures or conceptions of man are opposed to each other. The first one is a scientific-biological conception. The second one could be called a historicist picture of man.

In what follows I propose to defend the historicist conception.

(1.0) A traditional question is hidden behind the conflict of the biological-scientific and the historicist pictures of man. Is there a definite and permanent human nature manifesting itself more or less in all human individuals?

(1.1) From the affirmative answer follows a biological-scientific conception, for a permanent human nature cannot depend on historical chance. According to this conception, the bearer of the powers of mankind, including its intellectual, emotional and moral qualities, is mostly the genetical equipment. The individual features are, in a similar way, of biological origin: the given types of personality, for instance, are connected with physiological (eventually inherited) characteristics of the functioning of the nervous system.

A number of examples could be mentioned: theories of instinct, theories of aggression, characterologies, the theory of linguistic innatism, etc.

The approach outlined above is in general a reductionist one. This is reflected, on the level of philosophy of science, by reducing psychology to biology, *i.e.* by regarding psychology as a natural science. This is, for example, Chomsky's logic, according to which linguistics is a part of psychology conceived as a natural science.

It is possible to build an optimistic and a pessimistic conception on human nature. I hint again at Chomsky whose optimistic scientism is based upon a clearly formulated choice of value. He is committed to the idea of an invariant, biologically determined human nature, which is independent of contingent social-historical relations, because he believes that the opposite doctrine is favourable to reactionary social theories. If human beings had not an essential nature, if they were shaped or determined merely by the social environment, then – according to Chomsky – they would be subjected to unlimited manipulation, and we should be completely devoid of any moral ground to question the authorities and powers controlling them.

(1.2) What is called here a historicist picture of man can be summarized in the negative statement that there is no essential human nature. If there is a human essence or nature it is nothing else than the totality of social relations. The specific human properties can be defined only in terms of their relations to the forms and objects of human activity, but both the forms and objects of the activity take shape and evolve in the very activity and are not programmed in advance. What is human in human beings, and what is, therefore, the subject-matter of philosophical anthropology or human psychology, is merely a historical product, and cannot be identified with (cannot be reduced to) what is natural in them. As a consequence of this, the individual properties, and the personal characteristics are of social origin as well.

(1.3) The Chomskyan objections pose obvious difficulties for this conception. It is now enough to observe that, from the biological-scientific picture of man, just as from the historicist one, there may also follow reactionary and authoritarian social doctrines, but in the case of the historicist picture of man, as in the case of historicism in general, the relativistic implications are to be taken more seriously. I believe there is a kind of historicism in the framework of which it is possible to avoid extreme relativism, and the idea of the unity of mankind can be grounded as well as, or rather than, in the framework of the biological-scientific conceptions. Such a possibility is offered by the Hegelian-Marxian tradition.

(1.4) By reason of the above considerations, I propose to defend a historicist-Marxist picture of man and a theory of personality of Marxist origin.

(1.5) Incidentally, it must be observed that Marxism has its naturalistic version too. In particular, East European Marxism subscribed, for a long time, to a natural scientistic picture of man, conceived in the spirit of physiological reductionism. This was due to Pavlovism adopted by dialectical materialism. In some East European countries (mainly in Hungary and Poland, but in certain respects in the Soviet Union too), this kind of reductionist scientism has been supplanted in the past decades.

(2.0) Marx himself, in contrast with the naturalistic version of Marxism, traced back the characteristics of human activity, consciousness and psychic constitution to historical and social structures. It is in this spirit that he stressed that human nature is not an abstract entity inhabiting the individual person, but the totality of social relationships. He also stated that the formation of the five senses is the work of the whole history of the world.

There are two East European intellectual and scientific traditions which seem to have enriched the Marxian antinaturalism. The first one is the so-called Budapest School in Philosophy (with Lukács and his followers, like György Márkus, Ágnes Heller and others). The second one is the Soviet School of Psychology, represented by Vigotsky, Luria, Leontiev, and their pupils.

(2.1) The central notion of the philosophical anthropology of Lukács and his school is "species being" („Gattungswesen") of man. This concept refers to the fundamental fact that the bearer of the human powers is humanity as a species („Gattung"), contrary to the animal world where the capacities of the race are biologically determined and, as such, are given in every single individual. The constituting elements of "species being" („Gattungswesen") are sociality, consciousness, objectification, universality, and freedom.

Among these powers or constituting elements the leading role belongs to the factulty of objectification, which is a structural mark of labour. The fundamental activity of man is labour which results in an intentional and stable transformation of the environment. The objects brought about by labour are objectifications of subjective teleological representations, and of those skills and faculties which are indispensable for creating and using them. There is no subjective faculty which would not be embodied or materialized in some external object. This fundamental fact covers the totality of the human world: human beings objectify themselves not only in the physical products of labour but also in their institutions and habits, in their languages, scientific and artistic creations, in their ideas and philosophical systems, etc. And in all this are objectified not only their reason and intelligence, but also their senses, emotions, physical skills, etc.

"Sociality" as a constituting element of "species being" means that the essential human features cannot be deduced from what is given by nature (though the natural endowments are, of course, always presupposed). Social and historical life is a continuous process of pushing back natural barriers. This philosophical position excludes the notion of instincts held in common by animals and humans.

"Consciousness" as a power of "species being" means that, as regards the relationship between instinct and intelligence, intelligence has the leading role. The emergence of the human race is a story of the cutting down of instincts, for intelligence as a mechanism of adaptation to the external world supersedes the built-in regulation of actions.

The cutting down of instincts implies "universality" which means that man is an unspecialized being: anything may, in principle, become an object for his activity. In addition to this, the unlimited tendency of needs to extending and growing is also an essential aspect of man's universality. It is important that human powers in themselves, like labour, creative self-realization, knowledge, freedom, turn into needs.

The faculty of "freedom" is an openness to the future. The emergence of the human species is self-creation and, correspondingly, the formation of a person is a process of self-realization.

The anthropological conception, presented here, comes to be pertinent for a theory of personality through the mediation of certain assumptions concerning the relationship between the species and the individual. At this point, the following considerations arise.

(2.1.1) Without mastering certain minimal objectifications (language, customs, norms of behaviour) there is no way of becoming a human individual. This is an outcome of the sociality of man, and could also be expressed by saying that the development of personality is governed by social stimuli. It is, nevertheless, important to point out that what we are talking about is the appropriation of objectifications, and this does not amount to a simple learning process described in terms of stimulus-response or reinforcement-punishment. This has to be emphasized because behaviourism is not the unique alternative to the doctrines explaining behaviour in terms of innate mechanisms.

(2.1.2) As the human powers and faculties have their enduring existence in the shape of objectifications of the "species being" of man they may be alienated from the individual human beings. This is not only a possibility but also a basic fact of history. In other words, there is a discrepancy between the "species being" and the particularity of human individuals. It is from this space of alienation that arise those frustrations which explain the pathological distortions of the dynamics of personality.

(2.1.3) There is general agreement hat human individuals are singular, non-recurring beings and that this singularity, this non-recurring character is located in what we call personality. We have, however, a number of difficulties here. In the conceptual framework, elaborated by Lukács' school, certain difficulties are easy to handle. A distinction can be drawn between the particulartiy and the individuality of a human being.

The particular human being is, to put it in a simple way, the biological individual, who in possession of his primary needs, is a subject of everyday life. The term "individuality", on the other hand, refers to the "species being", *i.e.*, to that aspect of the individual person which is shaped by acquiring the most possible richness of the higher objectifications. This could be understood so that particularity refers to the distinct and autonomous character of a person, while individuality is marked by being absorbed by the social environment. But this interpretation is false. According to Lukács' teachings, the more conscious is the relationship of an individual to the higher values possessing the character of the "species being", the more he is a rich, original, autonomous, and "individual" person. The "normal" development of the personality can be described as a passage from mere particularity to individuality, as an ascent from particularity to the higher objectifications.

What is "normal" is not unconditionally the average. The alienated social relations confine most people to particularity. It can be said, once again, that a lot of psychic disturbances (if not most of them) are to be taken as symptoms of persons confined to mere particularity.

(2.2) The school of Vigotsky and his followers has been developing independently of Lukács' school but, in the frame work of a specialized branch of science, it constructed a similar conception.

I would lay particular stress on the view, formulated explicitly by Luria, according to which psychology is a historical science. This statement, referring to the status of psychology, corresponds, on the level of the theory of science, to the substantial thesis establishing the social-historical nature of psychic processes. As it is demonstrated by Luria's works, this conception does not neglect the scientific investigation of the neurophysiological bases of the functioning of the brain. It only claims the unity of physiological, psychological, and historical approaches.

How is it to be understood, in the conceptual framework of psychology, that the psychic processes and structures, constituting the personality, are of a social-historical nature? Vigotsky believed that the primary psycho-physiological phenomena (sensation and motion, attention and memory) are immediate, biological functions of the nerve tissue, while the psychic processes of higher order (intentional memory, active attention, abstract thinking, volitional activity) cannot be understood as immediate functions of the brain: their causes and origins are to be found outside of the organism. No doubt it is sensible to assume that in the hierarchy of the psychic processes the proportion of the immediately physiological factors and the external social ones is subject to change. However, the followers of Vigotsky, investigating visual and auditory experience, have made it evident that social determination must be taken into account on a more primary level, too. Because of the impact of linguistic structures on their organization, processes of sensation and perception have aspects which cannot be accounted for in merely physiological terms. (It is worth while noting the agreement of this position with Marx's philosophical insight regarding the formation of the five senses as a result of the whole history.)

The thesis on the social-historical determination of psychic constitution is opposed, in developmental psychology, to the conception according to which the psychic development of the child is a maturation of the forms, already given, of the psychic processes (that the development, therefore, is a simple quantitative growth which does not affect the structure of perception, attention, memory, and association). It is opposed, in general, to the view according to which the characteristics of perception and memory, of speech and thinking, the cognitive, emotional and volitional factors of personality are physiological processes which are essentially identical in any age and culture.

As is known, the school of Vigotsky has made important contributions to developmental psychology, cross-cultural researchers, psycho- and neurolinguistics, neurophysiology, and other disciplines. In developmental psychology, contrary to the view interpreting the psychic development as a maturing process, it established that the development of the child is characterized by a succession of distinct forms of activity, in the course of which it is primarily the very structure of psychic processes, rather than their content, which undergoes change. Great importance is to be attached to the cross-cultural research, accomplished in the thirties in Central Asia. This comparative research, proposed by Vigotsky and Luria, was designed to explore the dynamics of psychic processes under different social and historical conditions. As far as I know, this was the first important cross-cultural examination of psychic processes constituting personality. Apart from the results which strongly supported the principle of the sociality and historicity of personality I am referring to this research also because cross-cultural studies in personality came into fashion only in later decades and, by this reason, Vigotsky and Luria are to be regarded as pioneers in the field. Such a kind of cross-cultural approach to the dynamics of personality seems to be of a basic importance.

(3.0) I should like to indicate, in a very summarized way, what is implied, in different issues and fields of application, by the acceptance of the historicist picture of man. I shall touch upon the mind-body problem, the problem of communication, and the problem of psychiatry.

(3.1) The mind-body problem, *i.e.,* the problem of the relationship between the physico-physiological and psychic processes, is discussed, in general, in connection with cognitive states. This restriction is due to the Cartesian tradition which conceives the mind as a thinking substance. The actual literature, both in science and in philosophy, excludes the traditional, purely idealist solution to the problem and seems to permit two possible choices: dualism and a neo-materialist identity theory.

The identity theory can be summarized in the contention that mental states are identical with certain states of the brain. Subjective experiences, such as sensations and perceptions, correspond, factually, to definite physico-chemical states. Introspective reports, statements on one's own mental states are strictly speaking about physical processes in the brain of the speaker. The expressions used in neurophysiology and psychology have different meanings but, referentially, are identical. Being different names for the same things, they describe the same phenomena.

Perhaps the best known present-day form of dualism, as opposed to the neo-materialistic identity theory, is the conception of Popper and Eccles which I shall not treat here. I emphasize only that the historicist picture of man offers an adequate alternative against the dualist theory and the identity theory.

Putting aside arguments and counter-arguments, this alternative can be summarized in the two following statements: (a) the physiological and the psychic states and processes are distinct phenomena; (b) the psychic states presuppose, by logical necessity, the physiological states.

Statement (b) is sufficient for a materialist solution to the mind-body problem.

Statement (a) requires the following comments. The reason for distinguishing psychic states and physiological ones is that the former belong to the person, a social entity, and the latter are bodily states. The criteria of identity for a person, as a social entity, necessarily involve reference to interpersonal connections, while the criteria of identity in the case of a body can be given in pure physical terms of an observational language. If we consider the intentional character of psychic states, we have the same results. Psychic states, as intentional states, cannot be described without reference to an object of intentionality, while physiological states, and changes of states, can be identified in themselves.

Mind, as a totality of intentional representations, is realized in changes of state of physiological structures, and this does not contradict the contention that the psychological language cannot be reduced to a physiological language. A representation has a directedness, it is about something. The things, containing it, are about nothing. To put it in a simple way: what the neurosurgeon witnesses in the brain are neurons and he does not see the products of the functioning of neurons, experienced by us as perceptions and thoughts.

(3.2) Current linguistics and philosophy of language are faced by two contradictory hypotheses. According to the first one linguistic structures are previously given means of communication, in some way existing in themselves, which can be, therefore, described without referring to their use. In this sense they have priority over communication. The second hypothesis yields precedence to communication. In this sense the essential properties of linguistic structures depend on their use and cannot be described without examining the functions they have in the course of communication. The first hypothesis can be defended mostly in the framework of innatism, the second one involves the idea of the social character of language.

The historicist picture of man prefers the second hypothesis, connecting it with the thesis about the inseparability of language and higher psychic functions, like thinking. From the two theses – from the priority of communication and the mutual dependence of language and thought – it follows that higher forms of psychic activity are interiorizations of communicative actions. This can be generalized in view of the totality of a person. The personality is an interiorization of intersubjective relations, brought about, objectified and supported by communicative acts.

From this it follows that psychic disturbances and distortions of personality can be, partly or totally, interpreted as symptoms of distorted communication.

(3.3) As regards the issue of mental illness and psychiatry, it is not difficult to see that the historicist picture of man implies a stress on the social-communicative origin of psychic disorders and a commitment to a humanist psychiatry.

(3.3.1) Neuropsychic diseases, psychic consequences attributed to organic injuries and lesions, are to be treated apart. They belong, of course, to the competence of somatic medicine, and require, among others, the usual therapies, such as neurosurgical interventions and the like. The necessity to choose between the biological-scientist picture of man and the historicist one emerges as a relevant problem mostly in the case of mental disorders. The conflict of the two conceptions of man is translated here into the controversy of the biological (traditional medical) and the humanist model of psychiatry.

Humanist psychiatry rejects surgical interventions, such as lobotomies, on mental patients and questions the theories according to which mental diseases are determined by bioelectric or biochemical factors. This kind of criticism is not directed against the application of neurology, neurophysiology, or neurosurgery; it is concerned exclusively with the scientific, ethical and physiological foundations of psychiatry. It is not my duty to dwell on concrete arguments; what I wanted to call attention to was only what consequences do follow, in different issues, from a certain philosophical choice. The refuge of biological psychiatry is, actually, biochemistry. Most of those who contend that mental disorders, such as schizophrenia, are biologically determined take this view to be scientifically verified, although this is also a choice. The data can be interpreted in various ways. It is, for example, not known whether the biochemical concomitants of schizophrenia are causes or consequences of the illness, but it should be noted that the psychiatric labelling, in itself, is a question of social power, not to speak about criteria of normality and other difficult problems.

(3.3.2) Not all problems are solved by this social approach. The very concepts of mental illness and psychotherapy involve a contradiction, difficult to resolve, which is a reflection of a more profound contradiction between the natural or social causality and freedom, between the causal model of scientific explanation and moral autonomy.

While a theory of personality, by its very nature, lays a claim to scientific character, the personality, the structure and dynamics of which is described by the classificatory concepts of psychology, is a moral agent. It follows from the very concept of personality that a being to whom we do not ascribe a moral value is not considered a person. Persons are centres of moral autonomy.

In opposition to this, a scientific, psychological approach to personality, both in its theoretical and practical implications, has a reifying or objectifying effect: it reduces the sphere in which a person can be considered a responsible, autonomous and free individual. The very act of declaring somebody insane amounts to questioning the possibility of forming a moral judgement of him. Psychotherapy functions, in many respects, as an institutionalized system of declining responsibility.

This problem is raised not only by the authoritarian forms of psychotherapy. The unquestioned understanding and acceptance of the patient, professed by humanist psychiatrists, is in a certain way also paradoxical because it suspends criticism. Treating somebody as a moral being, or satisfying the needs of a person as a moral being, requires criticism just as acceptance.

I do not believe that any theory can answer these dilemmas definitely. The unique solution consists in learning, following Kant, to see human beings in two conceptual frameworks at the same time: in the framework of causality and freedom.

(4.0) By expounding the basic ideas of Lukács' and Vigotsky's school I did not want to give the impression that, in Eastern Europe, the conception of man prevailing in philosophy and in psychological and psychiatric practice is in a decisive way influenced by them. It would be difficult to tell whether there is at all a prevailing conception of man, characteristic of Eastern Europe. In any case, these two schools belong to the most significant philosophical and scientific traditions born in Eastern Europe. Their significance, in view of the problems treated here, consists in offering strong intellectual reasons for choosing the historicist picture of man and a theory of personality related to it.

References

Literature Regarding Vigotsky's School

Леонтев А Н, Проблемы развития психологии (Problems of psychological development). Издательство педагогических наук РСФСР, Москва, 1959

Lurija A R, Neuropsichologia e neurolinguistica, Editori Riuniti, Roma, 1974

Vigotsky I S, Мышление е речь (Thought and Speech). Издательство академии педагогических наук РСФСР, Москва, 1956

Vigotsky I S, Развитие высших психических функций (Development of higher psychological functions). Издательство академии педагогических наук РСФСР, Москва, 1960

Literature Regarding the "Budapest School"

Heller Á, La teoria dei bisogni in Marx, Feltrinelli, Milano, 1978[6]

Heller Á, Istinto e aggressività, Feltrinelli, Milano, 1978

Markus G, Marxizmus és antropológia (Marxism and Anthropology), Akadémiai Kiadó, Budapest, 1968

Lukács G, Die Eigenart des Ästhetischen, in Werke, vol 11—12, Luchterhand, Neuwied-Berlin, 1963

Lukács G, Zur Ontologie des gesellschaftlichen Seins, vol 13—14, Luchterhand, Neuwied-Berlin, 1971—1973

Correspondence: Professor Dr. J. Kelemen, Head of the Department of Philosophy, Faculty of Arts, Eötvös Loránd University, Budapest, Hungary.

Acta Neurochirurgica, Suppl. 44, 39–43 (1988)

The Concept of "Persona": Substance or Relation

G. Hottois

Institut de Philosophie, Université Libre de Bruxelles, Belgium

Summary

We oppose the substantialist to the relational conception of a "person" from a philosophical point of view.

Keywords: Bioethics; materialism; philosophy; person; personality relation; spiritualism; substantialism.

Introduction

The concept of "personality" is not a pure philosophical notion. It is rather a psychological concept. The notion of *person*, though marginal in traditional philosophy, has however begun during the 20th century to be endowed with a certain philosophical significance, even going to characterize what is commonly known as the Christian existantialism (the "personalism") in the middle of the century. But this concept which brings legal and ethical dimensions, became more and more significant in bioethics discussions which have increased during recent years.

For the philosopher—and I am presently speaking as a philosopher—it is the concept of person which is important, the "personality" being considered as both a peculiar and an individual determination of a human being.

In this lecture I would like very simply to show the evolution of the concept of person from a substantialist conception (metaphysical or theological) to a relational conception which is a contemporary notion.

1. The Substantialist Conception

In fact it corresponds to the classical metaphysical conception, which also applies to a large extent to the theological point of view. According to this concept, the person is a subject and this one is substantial.

What is the philosophical meaning of a substance? The answer obviously differs from one philosopher to another and the word does not mean the same for Aristotle, Descartes or Locke. Nevertheless it is possible to point out some characteristics which are more or less constant.

The substance is a basic reality, irreducible, therefore elementary and simple; it is self-supporting, that means that it is autonomous and does not need anything else to be existing. Being simple and fundamental, it is both non-evolutive and immutable and should it have a relationship with other substances it cannot affect its essential character. For instance, Descartes distinguishes between two substances: the extensive (the matter) and the thinking one (the spirit). The matter is inert; the transformations that it undergoes are mechanical; its shape and performances could be computed.

On the other hand, the spiritual substance is endowed with thought and consciousness; it perceives itself, knows itself thanks to the reflexive process. The most radical expression of this inner consciousness is well known, as the "cogito ergo sum". Endowed with thought, conscience, reflexiveness, the spiritual substance regards itself everytime as a subject. All that is not subject, that means the whole lot of objects, is related to the other substance, the physical one.

The subject is spirit, soul, thought, mind. It is free, autonomous. That means that it cannot be affected either by causal sequence, or by the technical manipulation characteristic of the physical world, corporeal and natural. By itself, and in its deepest sense, the spiritual subject is also indestructible (immortality of the soul) because it is simple. Finally, it is universal. The subject being characterized by the thought, the thought itself being rational and reason being the same for all the subjects.

It is why Descartes did not valorize the dialogue, that is the communication. Being, as everyone, bearer of the universal consciousness, he was assuming that he bore within himself all the necessary light for a total rational knowledge.

What place could have the concepts of person and personality in this frame of metaphysical substantialism? When one said that every human being is a person and that we must respect this person as he is, one is not very far from the idea of subject, characterized by its autonomy, freedom, thought, its non-reductibility to a material thing which could be manipulated, destroyed, possessed. However, even considered from that standpoint, the idea of person *desubstantializes* the concept of subject, that is to say the human person has to be protected. It is not undestructible or immortal as the soul is. It has to be protected because it is of value, and not because it is related to any other supranatural world. This reasoning is especially appreciated in the field of bioethics. For example, with regard to the status of the embryo, those who are in favour of prohibition of any kind of experiment or manipulation on it, no longer state that the embryo, as soon as it has been conceived, is endowed with a soul (the metaphysical subject). They will argue that the embryo has to be respected as a person, or even more unpretentiously, as a *potential* person.

In other words, we somehow proceed, more or less clearly, from a concept of a reality which is of value because it proceeds from another nature (spiritual, divine), that means it bears value by itself independently of future or any acts of valorization, to the concept of a reality which is of value because it could be invested with some meaning, with acts of valorization under some prevailing circumstances. I shall deal with this point again (v.i.).

As far as the concept of personality is concerned, I would say that it does not have much place in the frame of metaphysics. Besides, its description, under the name of character, belonged in the past more to a literary style than to a philosophical one, the mortalist's style. In fact, trying to place the concept of personality in the frame of the substantialist metaphysics directly leads to one of the "aporias" of the classical metaphysics, that of dualism. That was the unresolved problem that Descartes faced: how two substances—the spirit and matter—totally heterogeneous could interact, as they are doing, to all appearances, in the human being who is a mixture composed of one body and one soul.

Fundamentally, if we try to find a place for the personality—which varies from one human being to another and which evolves during the lifetime of a single subject—in the frame of metaphysics, this place can be but the compound itself, the medium which at the same time separates and joins together the two substances. The personality is the result, each time different, of the incarnation, each time individual, of the pure subject in an impure body or in the vile and contemptible matter. The classical philosophy had two reasons to stay beyond: first its failure to explain the relation between the body and the soul; secondly its disinterestedness, even its contempt for all that is related to the body, to any material, empirical description. The philosophical psychology was idealistic: it was theory of the soul, that means a pure subject, not tarnished by all that proceeds from a bodily nature.

So has been briefly conveyed the philosophical thought which was predominant until the 19th Century (and which still has various prolongations during our era).

I would now like to discuss briefly some aspects of the criticism of the substantialist pattern before going on, as a conclusion, with the evocation of what I would call the relational and dynamic pattern.

2. The Criticism of the Substantialist Pattern

Once more, I am limiting myself to the criticism of philosophical origin, especially of a French-German origin*, since it is in Europe that metaphysical substantialism has been so far the most powerful.

This criticism has mainly arisen from the phenomenological movement (Heidegger, Merleau-Ponty). The phenomenology has criticized the simple and plain distinctions of metaphysics: subject-object, body-soul, matter and spirit, language-thought, self-thou, inner reality-perceived reality, etc. From this standpoint what finally does exist, is the intricacy, the complex and moving binding of these opposed entities. Such a reality is at the same time global, structured, relational, complex and dynamic.

There is not on the one hand the human being and on the other hand the world, but there is the human being within the world and its various modalities. The same applies for the body and the spirit. What does exist is an intricate psycho-chemical structure, not reductible, which expresses itself in various manners and in an evolutive way. There is neither juxtaposition, nor

---

* In England, the situation is quite different on account of the powerful empirical movement which spreads out as early as the 16th and 17th centuries. Nevertheless, from the global point of view we have here adopted, the criticism made by the analytical philosophy (Rawls e.g.) to the metaphysical substantialism comes upon, with regard to many aspects—although expressed differently—the phenomenological criticism.

later communication of a number of isolated subjects: from the beginning onwards there is the being with the other, that is the constitutive intersubjectivity of individual subjects; there is from the origin, not the monologue, but the interlocution and the dialogue. In short, the pure subject, monadic, does not exist, it is a metaphysico-theological illusion, the subject is always involved in a complex and dynamic network of relationships which are building him up ... Even his relationship within himself becomes complex and impure. There is no more the timeless *cogito* but for everyone a relation with his being which is relational and projective (always in being to come). Phenomenology has set precisely in the centre of the philosophical attention what the classical metaphysics did not want to know, viz. the relation, the complexity, the mixture, the structural, the dynamism. In some way, it is therefore in a perfect harmony with contemporary science which faces the evolving complexity. However the phenomenological denial of the analytic and objective thought, considered not without reason as overthrowing the apprehension of the complexity and the relationship, most often results in a clash between the phenomenologic philosophy and the technico-scientific undertaking of knowledge and mastery of the complexity.

Still it remains that phenomenology has contributed to improve the conceptual tools capable of giving a new meaning to the concepts of person and personality.

3. Aspects of a Relational Conception

It is in the frame of medical ethics and especially of bioethics that the opposition between the substantialist conception and the relational conception of the person has recently been recognized as a preponderant position, but this position is not a simple one.

The Ambiguous Persistancy of Substantialism

It is obvious that the moving back of the traditional conception of a pure and spiritual subject is due to a great extent to the development of the modern science and its positivist methodology. However this methodology has also given rise to a kind of metaphysics, namely materialism. In this trend a concept of the person (and of the personality) gained strength and finally expressed itself in plain words of physical or physico-chemical meaning according to which, as a whole, the spirit would only represent the complex result of cerebral activity, and the essential being would become identified, all things considered, with the human genome. In that manner the human being happens to be, presumably, totally objective (or objectivable) and, in theory, exposed as any material object, to any kind of manipulation or experimentation.

I would only like to emphasize three remarks with regard to this trend of opinion. Firstly, this theory coincides with a substantialism which is the reverse of the formerly prevailing metaphysics. In this sense, it is simplistic and excessive. It is not because, up to now, the causal materialism as a method, has allowed such considerable progress that it is rightful to transpose it into a concept of the world which would exhaust the meaning of reality, including the human being. Secondly, this materialist metaphysics inevitably gives rise to a renewal of the opposite and adverse conception, above all when the human being is concerned: the spiritualism, as the recent positions adopted by the Vatican regarding human procreation have demonstrated it. Finally, no more than the spiritualism, does the materialism—since it is also a substantialism—treat the relational conception of the human as it deserves. And should it refuse to deal with the human being as a pure object, it is still unable either to tell, or to justify or to grant the reason why it declines to do so.

Towards a Genuine Relational Conception

Going through a quite recent monography on bioethics (Brody and Engelhardt, 1987) in which the concept of person is mentioned, one finds the following determinants of the specificity of the human being:—consciousness (relationship to oneself), interest (desires, propensity towards objects), language, sensibility bound to affectivity (sentience). These various characteristics are set up by the authors not only as criteria of the person but in the same time as a fact that the human being to whom they are applied deserves respect and has rights.

All these criteria are relational by definition (most particularly the last three: language, interest, "sentience"). When considered as a person the human being is made of a network of relations, in which he is developing and whose development he contributes to. This network of relations is carrier of meaning and values. For the relations which are concerned have no causal character: they are symbolic or significant; they settle priorities and hierarchies.

This relational linguistico-affective nature of the person assumes three forms:

— the person is a result of linguistico-affective relations (the new-born only becomes a person in an adequately relational environment),

— the person is producing relations,

— the person is object of relations (caught in a network of linguistico-affective relations issued from other people).

This is obviously true for the adult as well as for the child, for healthy people as well as for sick ones, but this is also true with regard to the foetus: quite naturally, from the very moment of his conception, he is already engaged in the linguistico-affective network of his parents, his family, hence the legitimacy of the concept of potential person. That could also be true for the individual sunk into a "coma dépassé" providing that some people continue to consider him as an object of relations (for instance through the affection of parents toward an injured child, who continue to invest themselves affectively into him). Hence the legitimity of the notion of residual person. But—and this is essential—this relation character of the person also makes it survival or its symbolic existence *to a certain extent* depending on the context, thereby, providing its practical effects resulting from the difference with the substantialist conception.

It is obvious that a non-desired foetus, indeed rejected by its mother, is not considered by her as a person and that the relational environment in which it could be born, would not be propitious to the development of its personality (the quality of life, which is also as intrinsically relational notion).

It is still more obvious that supernumerancy foetuses are not considered as potential persons and that the relational environment in which they have been conceived is completely different from the notion normally attached to a conception.

Finally, it is manifest that the individual in a "coma dépassé" and whom nobody is taking care of is no more relationally involved as a residual person. In such circumstances, it does not appear immoral that foetus or vegetative people be involved in another relational network which is not the family any more, the society providing however that they are not treated as pure objects, because whatever is the attenuation of the quality of the person (potential or residual or fully marginal, the psychotic or deep backwardness) this never completely disappears. Symbolically and not only genetically the supernumerary foetus, as well as the vegetative people, somewhere has a tie with mankind. Therefore if it appears legitimate to make use of them in a therapeutic research project (or as donor of organs) it does no seem legitimate to make anything out of them.

Thus the experimentation on the embryo with a medical aim, or the use of an aborted foetus as donor of organs does not constitue a plain and simple objectivation of the human being, but a reinvolvement of the latter in a social and humanitarian project, taking into account the context pertaining to each case.

The substantialist conception of the person, spiritualist or materialist, does not allow such a flexibility, guarantor of freedom and respect. This s why it leads to abuses and contradictions. Although I have not approached other aspects of this problem due to lack of competence, I am of the opinion that this linguistico-affective relational conception of the person ought also to be valid for all the problems of bioethics when we are confronted with severe disorders of the personality, disturbing the relational capabilities of the subject and therefore his status as a person.

Lastly I would like to come to a conclusion by suggesting that this relational conception may be—unlike the substantialist conceptions—universalized without danger because of its respect of the freedom and of the difference. The one who adopts it may accept that others claim a spiritualist conception and refuse for example abortion and insemination *in vitro*. The Christian is completely free, from the relational point of view, to consider that a child heavily affected by genetic disturbances has the right to be born. However one may expect that the family or the Christian community be capable to assume the choice to its very end, in full coherence, by carefully looking after the desired child and not by shifting the responsibility of this care on a pluralist community of which certain members would not have made such a choice, being maybe more aware of the human limits in the capacity for affection and assistance.

But in return, what could not be accepted is that the followers of the spiritualism want to dictate their substantialist concept to those people who are not of their opinion. Now, we have to say that the substantialist conception, being metaphysical, is dogmatic by its own logic.

However, the adept of the relational concept of the person principly speaking cannot agree with the integral materialism which, objectifying grossly the human being, leads to an unethical nihilism, denying the human.

As a matter of fact, there are nowadays, many materialists but very few of them are going on with lucidity up to all the logical and philosophical consequences of their affirmation. In such a way that the materialist in general is not at all inhuman but simply involved in deep contradictions that he does not even perceive. Nevertheless, it remains that dangerous materialistic temptations (technicists) exist and that as such they deserve, inasmuch as the tendencies to the dogmatism, the vigilance from those who are anxious to preserve a pluralist community. The relational concept of the person (and of personality) is, to my mind, a condition of survival for such a community.

As the attentive reader will have observed it, the relational conception, schematically developed here, is expressing the laic point of view of the Free University of Brussels. This conception refuses as much the objective and the relative materialism or technicism, as the spiritualistic, metaphysical or religious dogmatism; both can lead to abuses in which the concept of human person itself is failing. In addition, in this lecture, I have particularly favoured the relational conception because it won a great actuality in the western philosophy unlike the substantialist conception, still current but old and better known.

References

1. Brody BA, Engelhardt HT (eds)(1987) Bioethics, reading and cases. Prentice Hall, Inc, Englewood Cliffs, New Jersey
2. Hottois G, Susanne C (1988) Bioéthique et libre-examen. Ed de l'Univ de Bruxelles
3. Humber JM, Almeder RI (1976) Biomedical ethics and the law. Plenum Press, New York
4. Lemaire J (ed) (1986) Naissance, vie, mort: quelles libertés? Ed de l'Univ de Bruxelles;
5. Ribes B (1978) Biologie et éthique. Paris, UNESCO

Correspondence: Prof. Dr. G. Hottois, Institut de Philosophie, Université Libre de Bruxelles, Bruxelles, Belgium.

III. Methodology of Personality Evaluation

Acta Neurochirurgica, Suppl. 44, 47 (1988)

Introduction

G. Stroobandt

Brussels, Belgium

A session devoted to personality and neurosurgery will concern the disturbances of personality in neurosurgical diseases and after neurosurgical interventions. Most neurosurgeons would be embarrassed if they had to give a definition of "personality", because they are neither psychiatrists nor philosophers. But every neurosurgeon and, even, every house doctor knows perfectly well what *disturbances* of personality mean. All of them have experienced, in their medical practice, that neurological impairment was not necessarily limited to obvious deficiencies like hemiparesis, aphasia etc. Patients who sustained major brain damage followed by recovery, may nevertheless remain afflicted by changes of character with intellectual and mental sequelae that considerably change their "personality". Let us take the example of a brilliant student or of an eminent scientist, who ultimately recovered from a severe cranio-cerebral injury, but remained merely the shadow of his former self. Let us also consider the example of a man who recovered from a severe subarachnoid haemorrhage caused by the rupture of an anterior communicating aneurysm and was successfully operated upon. Some time later, he looks in perfect condition, but his relatives may complain that his character, his moral sense, in a word, his "personality" has changed, as a consequence of the disease. Although nowadays aneurysm surgery gives much better results than some years ago, it still cannot avoid either a small amount of postoperative mortality and morbidity, or, above all, the consequences of the bleeding itself.

Lesions in the frontal medial and basal areas, disruption of ridiculously small vessels in that region may result in dramatic situations such as we are confronted with in no other fields of surgery. A correct appreciation of such disorders is therefore essential in making decisions about operative indications, technique and strategy, relying upon personal experience and literature data. But it becomes then absolutely necessary to speak the same language and to give the words we use, the same sense. It is, for instance, amazing what differences may exist between the meanings of words like "good" and "fair" in characterizing the results of aneurysm operations, according to different authors. Säveland *et al.*[1] (1986), checking their patients after aneurysmal rupture and early operation, concluded that persisting cognitive disturbances and psychosocial impairment should be taken into account in any outcome assessment; if such incapacity was severe, the outcome should be considered unfavourable, even in individuals without major neurological deficits.

We should have to complete our neurosurgical outcome evaluation by a personality outcome scale, using terms that take on the same connotations for all of us, beyond our own personalities and cultures. Such an enterprise might be difficult, but it could be worthwhile.

References

1. Säveland H, Sonesson B, Ljunggren B, Brandt L, Uski T, Zygmunt S, Hindfelt B (1986) Outcome evaluation following subarachnoid hemorrhage. J Neurosurg 64: 191–196

Correspondence: Prof. Dr. G Stroobandt, Square Vergote 12-21, B-1200 Bruxelles, Belgium.

Acta Neurochirurgica, Suppl. 44, 48–53 (1988)

Methodology of the Evaluation of the Personality

W. Huber

Department of Psychology, University of Louvain, Belgium

Summary and Keywords see page 53

I. Introduction

When I was beginning my work in a neuro-psychiatric hospital, a senior physician one day told me: "You see, our problem here is that we have three types of colleagues: First we have neurologists who have a great knowledge of the nervous system, but can do relatively little for their patients, then we have surgeons who don't have so much knowledge, but who are very effective, and finally we have psychiatrists who don't know very much and are not very effective". As the psychology of personality is closer to psychiatry than it is to neurology, you may imagine the apprehension I had when asked to give a paper on personality before an audience which combines knowledge and efficacy! But, doing my duty, I came to realize that, when respecting certain methodological requirements, we could know and perform reasonably well.

The object and the domain of neurology are quite clearly defined and there is quite substantial agreement here. Unfortunately this is not the case with the psychology of personality, and for this reason it is important to outline some of its issues in order to discuss successfully the present situation of the methodology of the evaluation of the personality.

In order to do so, I would like to remind you first that the psychology of personality conceives of itself as a fundamental discipline, which aims at describing, explaining, and predicting psychological structures and processes, and thus provides the knowledge necessary for the clinician, be he a therapist or an educator, who wants to make an intervention in order to help a client the better to develop his personality or to solve his problems. The psychology of personality conceives of itself then as the empirical discipline, which means as a discipline making its observations in the context of a theory and in controlled conditions, and making its inferences according to principles and procedures of inferences prevailing in the empirical sciences.

The object of this discipline, *i.e.* personality, has been defined in different ways, but for our purpose we could clarify things by distinguishing two types of definition. First a narrow one, and second a broad one.

The narrow definitions which belong to the older group of definitions conceive of personality as the core of the person, comprehending primarily affectivity, the believes and attitudes in the domain of values. The broader definitions, which are currently prevailing include, in addition to affectivity, beliefs and attitudes about values, such domains as morphology, physiology, temperament, needs, interests, attitudes and aptitudes. In such a definition you could say that personality is "a person's unique pattern of traits" (Guilford 1959, p. 5) and that a trait is "any distinguishable" relatively enduring way permitting to distinguish one individual from another" (ibid., p. 6).

II. Major Issues in the Psychology of Personality

Having given a first general definition of the psychology of personality and of its object which is the normal as well as the abnormal personality, let us have a look at some of its major issues that also have a bearing on the problem of evaluation.

First there is the question: idiographic or nomothetic approach? or, in order to study personality, is it better to investigate an individual as intensively as possible, even when this should not allow you to generalize your observations to other individuals, or is it preferable to try to establish general laws, which normally requires the observation of many individuals and may lead you to sacrifice the originality of the individual? The answer is that both approaches are necessary. It is not possible to explain the individual behaviour without reference to empirical generalizations and these cannot be secured by the idiographic approach which does not provide the necessary conditions of control and repeatability.

A second issue is then: Categorical or dimensional systems? Here the question is to know whether personality data should be organized along categories and types, or along dimensions. Should personality be conceived of and organized as a series of dimensional characteristics that combine to give a unique profile for each person, or should it be conceived of as a typical pattern of certain qualitative characteristics?

The third problem, a big one, is the problem of stability and consistency versus situational specificity. Is the behaviour of an individual enduring, stable over time, and consistent across situations, or is this an illusion and is behaviour primarily determined by the situation in which it occurs?

This debate, already investigated by Hartshorne and May (1928), has re-emerged with a book by W. Mischel in 1968. The present state (s. Endler and Magnusson 1976, Epstein 1979) could be summarized as follows:

1) Individuals differ in the degree to which their behaviour is stable and consistent.

2) An individual is stable and consistent in certain characteristics only and not in others.

3) Each individual is stable and consistent in only a small number of characteristics which are those he experiences as being central to his personality.

This relatively small number of stable and consistent traits is what we generally mean when speaking of personality.

As to the determinants of the behaviour, the current position is that the behaviour is a function not only of the situation or of the personality, but that it is a function of the interaction of the personality and the situation.

A fourth issue pertains to the conceptualization and analysis of structure and process, of structure and change, and their relative importance for a psychology of personality. What and which are these relatively stable and consistent elements, what are their relationships, their organization? How to conceptualize and analzye process, change and development, which are important phenomena in a psychology of personality? To these questions different answers have been given as may be seen, for example, in the theories of Freud, Allport, Cattell, Eysenck, or Mischel.

A last big issue then is the issue nature-nurture. Concerning the respective role and the importance of biological and social determinants of behaviour and personality, its change and its development, normal as well as abnormal, recent medical, genetic, physiological, and developmental research has considerably enlarged our knowledge about the biological factors, about their effects on behaviour and personality, and about their mechanisms (*e.g.* Bradford 1987, Mendlewicz 1987). Nevertheless we should not forget that, despite these advances, for the moment, we do not seem to know of simple physiological or biochemical indicators of the dimensions of personality (Fahrenberg 1985, p. 107). We should also say that concerning the development of behaviour and personality disorders, the so-called medical model and the psychosocial model, being too simple, are being replaced by the biopsychosocial model which poses the hypothesis of the multifactorial pathogenesis and conceives of these factors as being in constant interaction (*e.g.* Weiner 1978).

Having thus been reminded of the major issues we should point to the fact that the aim of the psychology of personality is to elaborate systems that permit the description of a large number of all individuals, and the explanation of the inter- and intraindividual differences. Its aim is not to study the individual personality, this is the object of clinical psychology and psychodiagnosis.

III. Major Questions of the Assessment of the Individual Personality

To assess an individual personality implies answering a series of preliminary questions we should be aware of if we want to get a maximum out of a procedure and not to expect information which this procedure cannot give us. Some of the major questions then are the following.

First: Classification or assessment? This means: Do you want to classify an individual case by putting it into a classification system, nosological or dimensional, for instance, or do you want to go beyond and to obtain knowledge about its individual dynamics, the interplay of the assets and deficits that determine this individuals behaviour and its evolution?

Second: Status or change? Do you want to evaluate the present state or the changes in this personality, these changes being a product of spontaneous recovery or deterioration, of a pharmacological or a psychological intervention?

Third: Do you want to make a norm oriented or a criterion oriented assessment? Is your aim to evaluate personality by comparing the case with respect to norms coming from a representative sample, or by comparing it with respect to individually set criteria as, for instance, a certain level of individual functioning to be obtained after therapy?

Fourth: Tests or inventories? Would you assess an individual on certain variables measured by a certain test, anxiety, aggressiveness, for instance, or do you want an inventory, a listing of all the characteristics pertaining to a certain problem. For instance instead of wanting to know to what degree X is anxious or aggressive, you may be interested in the situations eliciting his anxiety or agressiveness and make an inventory of these situations.

Fifth: Measurement or information gathering? Here the question is: Are you interested in having a measure of a characteristic, a degree of depression for instance, or in having information useful for treatment decision and planning? And, it should perhaps be added that this latter information does not always need to be highly quantified.

IV. Sources and Methods for Gathering Data in Personality Assessment

Bearing these questions in mind, what then are the sources and methods for gathering the data in personality assessment? There are different classifications of the sources and types of data and of the methods allowing one to gather these data, depending on various criteria of form and of content. For our purpose let us remember the distinction made already by Wundt who distinguished self observation versus hetero-observation, and pure observation versus controlled observation. And let us remember also a more recent distinction proposed by Webb *et al.* (1966) distinguishing reactive data and non-reactive data, reactive data being produced instrumentally and systematically, whereas non-reactive data are spontaneous data. With reference to these distinctions we could then classify the different methods of gathering data by locating them in a two-dimensional schema, one dimension referring to the observer, the other to the mode of production of the data.

1. First you could have gathering of *non-reactive data by self-observation.* Here, the raw material is spontaneous productions of people, what people say spontaneously about themselves, as for example, letters, diaries, conversation notes, therapy session records, and so on. The elaboration procedures for such data are 1) interpretation and 2) content analysis (systematic and quantitative analysis of linguistic productions).

2. A second possibility is the gathering of *non-reactive data by hetero-observation.* The raw material here is behaviour and its traces. The procedures of elaboration are: 1) the direct or indirect observation of behaviour. This requires special training and also adequate observation systems as well as the awareness of possible interaction between observer and observee; 2) the analysis of documents, of the traces of the behaviour. Here you use principally interpretation, content analysis, or simply computation of frequency.

3. A third possibility is the securing of *reactive data by self observation.* Here the procedures are first the questionnaires which are of different types. You can also have different classifications, but we could say that in function of the number of variables studied you may distinguish univariate and multivariate questionnaires. In function of the type of question and answer you may distinguish open, semi-open, and closed questions. You may further distinguish rating items versus forced choice questions. It is important also to say here that the wording of the question is very important as has been shown by recent research. Another possibility of elaboration of these data is then to use the semantic differential, to take polarity profiles, and finally, another possibility is to make a Q-sort.

4. A fourth and last group of possibilities is to secure *reactive data by hetero-observation.* The raw materials here are the traces of what people do or say in controlled situations as seen by someone else. The procedures are first objective tests such as those proposed by Cattell or by Eysenck, and then the interview. Here you have to distinguish the free interview, the semi-structured interview, and the standardized interview. These latter distinctions are important as has been shown when the researchers and the clinicians were elaborating DSM III categories and the DSM III diagnostic system.

Three remarks should be made here. First, it is important to be aware that the results secured by all these procedures may be influenced to a more or less important degree by systematic and non-systematic errors having their origin in the patient and in the assessor. In the patient for instance, there are the thoughts, the expectations, implicit theories of personality, simulation, the response styles, etc. In the assessor you have also the implicit theories of personality and the fundamental error, that means our tendency to give more importance to the inherent qualities of personality than to the determinants of the situation. You have then also the different judgmental errors, such as the halo effect, the leniency effect, the logical error, etc.

Secondly it is important to know the methodological

qualities of these different procedures. This means the objectivity, reliability, sensitivity, validity and utility (what information an instrument gives you beyond what you already know).

Thirdly, the results obtained for a given individual may vary and diverge as a function of type and sort of data, as a function of the instruments used, as a function of the evaluator and the moment of the assessment.

The preceding characteristics and remarks then confront us with the problem of choice. What procedures should we adopt? The answer is that nowadays we have at our disposal a great number of procedures and instruments of quite variable value. We have to choose and to combine them accordingly to the problem we have to solve, for instance classification, patient selection, therapy planning, assessment of change, etc. This has to be done by choosing and combining in a rational way the procedures and instruments for which we know the value and utility for solving a particular problem. As a general rule we could say: "Be specific, but don't lose sight of the problem". In her discussion, Dr. Sichez-Auclair (see below) gives an excellent illustration of some of these methodological problems encountered in the case of brain damage and personality.

V. The Process of Clinical Judgement

This takes us then to a last problem: the process of clinical judgement or the clinical inference process. This is the process by which the data collected according to different questions are elaborated and integrated to form a final assessment. It may be conceived of as a problem-solving and a decision-making process that occurs in a social situation.

Here we should point to the fact that this is a sequential process of investigation and verification where the different moments are interactive, and its final product is a synthesis of the problem that has to be communicated to the person who asked the referral question. In order to better understand this very complex process, let us distinguish its phases, the factors of influence, the styles, and ask the question of clinical or statistical prediction.

1. Phases of the Clinical Judgement Process

a) Schematically we could say that at the first moment of the assessment situation a first impression grows more or less consciously and influences the formation of the initial hypotheses.

b) At the second moment, a strategy and method of investigation emerges and is implemented to secure data concerning the problem.

c) In the third phase these data are progressively organized, and a final synthesis describing and explaining the problem and permitting one to make predictions and decisions is elaborated.

d) At the last stage, this final assessment has to be formulated and communicated adequately to the person asking the referral question.

2. Factors Influencing Clinical Judgement

What are the factors that influence this process of combining and elaborating the different data in personality assessment?

a) There are the factors of the institution and of the situation of assessment. Here it is important to know that the position of the diagnostician, the moment of assessment may be important as shown for example by the research done in pain clinics.

b) The theoretical orientation of the assessor is also very important because it will have him look for and see different data and also elicit different reactions from the patient. His theoretical orientation will also influence the way of looking at data and the level of inference.

As to the way of looking at data for instance you may look at data as *samples* of behaviour, you look at what the patient does in a specific situation. To be valid, such observations must also give the frequency and the conditions of manifestation of the behaviour. Or you may look at the data as *correlates*. You look for other characteristics that are or could be associated with what you observe. This is the procedure preferred by psychologists of the psychometric orientation. Here, to be valid, a judgement requires that you know the tests and the norms. And finally, still another way to look at the data is to conceive of them as *signs*. The observational data are then conceived of as signs for underlying conditions or traits. This is the psychoanalytic approach and it is important to say that the validity of such judgements requires that you check the validity of the theoretical proposition and also the validity of the individual interpretation.

As to the level of inference then, you may have low, medium and high levels, following the degree of abstraction and complexity of your inference.

3. Styles of Clinical Judgement

Finally something should be said on the style of clinical judgement which is another factor that influences the end product.

a) First you have what traditionally is called the clinical style, or what Sundberg and Tyler (1962) call, very happily I find, the way of emergent synthesis. Here you are in the idiographic tradition, which aims at knowing the person, the gestalt, in its totality and its unity. The mental processes intervening are empathy and intuition, and intuition is conceived of not as a syllogistic reasoning but as a comprehension of the significance of the behaviour, thanks to a comparison with prototypes and thanks to a feeling of evidence when there is a real correspondence.

b) Another style is what Sarbin *et al.* (1960) call taxonomic sorting or, if you prefer, syllogistic reasoning. Here the different elements of the judgemental process, the cognitions, facts, feeling, etc. are related to each other and integrated following explicit and conscious rules, and the conclusions are arrived at by syllogistic reasoning. Here you try to understand the individual not by comparing him with a prototype but with reference to general laws, and you do not aim at understanding him as a totality or as a unique person, but you only want to know certain aspects of his personality.

It is important to point out that, first, in both approaches you need to verify correctly the hypotheses and that this is not a matter of intuition. And second, these approaches are not mutually exclusive, but they are processes the importance of which varies following the moment of the diagnostic process, for instance hypothesis generation or hypothesis verification, and the importance varies also with clinicians.

4. Clinical or Statistical Prediction?

The question is here to know: to make correct predictions or assessments, is it better to elaborate and to combine data following the clinical way, following your intuition and clinical experience, or is it preferable to adopt a syllogistic and statistical procedure which combines data following pre-established rules and formula? This question has been asked by P. Meehl in 1954 and has given rise to a hot debate and to quite a lot of research (*e.g.* Holt 1970, 1975). The present state of this debate might be summarized as follows (*e.g.* Holt 1970, 1975):

1) There are individual differences in the capacity for making clinical judgements and predictions, some clinicians doing much better than others.

2) The analytic and disciplined judgement is better than the global and intuitive judgement.

3) The prediction of a behaviour is better when the evaluation of the personality is complemented with an evaluation of the situation.

4) Where there are statistical procedures available, the clinician should use them. They are not always available and often not at all.

5) And finally, the quality of a statistical judgement and of a computer assisted judgement depends on the quality of the initial clinical observation.

With respect to this last statement it is interesting to read a sentence by Wiggins written in 1973: "In the realm of clinical observation and hypothesis formation, the IBM machine will never be more than a second rate clinician" (Wiggins 1973, p. 199). Ten years later, Spitzer (1984) has the following answer to the question "Are clinicians still necessary for psychiatric diagnosis?": "the burden of proof is now on the clinician to show that the advances in technology have not made the clinician superfluous in the task of diagnostic assessment. Whatever the outcome of this game, rematch is inevitable" (Spitzer 1984, p. 287).

I think one can certainly subscribe to this sentence, but without forgetting that the personality assessment goes beyond psychiatric diagnosis.

To conclude, we may say that, despite its problems and difficulties, personality assessment is still worthwhile and that research demonstrates that when it is done following "les règles de l'art", it is interesting and useful not only for the researcher, but also for the clinician (as shown *e.g.* by the work of Brooks 1984, and Sichez-Auclair 1984) and that it may still be improved.

References

1. Bradford LD (1987) Pharmacology and psychotherapy in affective disorders. In: Huber W (ed) Progress in psychotherapy research. Selected papers from the 2nd European Conference on Psychotherapy Research. Sept. 3–7, 1985, Presses Universitaires de Louvain
2. Brooks N (1984) Closed head injury: psychosocial, social, family consequences. Oxford, University Press
3. Cattell RB (1965) The scientific analysis of personality. Harmondsworth, Middlesex (Penguin Books)
4. Endler NS, Magnusson D (1976) Toward an interactional psychology of personality. Psychological Bulletin 33: 956–974
5. Epstein S (1979) The stability of behaviour: I. On predicting most of the people much of the time. Journal of Personality and Social Psychology 37: 1097–1126
6. Eysenck HJ (1960) The structure of human personality. Routledge and Kegan Paul, London
7. Fahrenberg J (1985) Biologische Grundlagen. In: Herrmann Th, Lantermann E-D (eds) Persönlichkeitspsychologie. Urban und Schwarzenberg, München, pp 101–110
8. Guilford JP (1959) Personality. McGraw Hill, New York

9. Hartshorne H, May MA (1928) Studies in the nature of character. Vol I. Studies in deceit. Macmillan, New York
10. Holt RR (1970) Yet another look at clinical and statistical prediction: Or, is clinical psychology worth-while? American Psychologist 5: 337–349
11. Holt RR (1975) Clinical and statistical measurement and prediction: How not to survey its literature. JSAS Catalog of Selected Documents in Psychology 5: 178, MS n° 837
12. Meehl PE (1954) Clinical vs statistical prediction. A theoretical analysis and a review of the evidence. University of Minnesota Press, Minneapolis
13. Mendlewicz J (1987) Manuel de psychiatrie biologique. Masson, Paris
14. Mischel W (1968) Personality assessment. Wiley, New York
15. Sarbin TR, Taft R, Bailey DE (1960) Clinical inference and cognitive theory. Holt, Rinehart and Winston, New York
16. Sichez-Auclair N, Sichez JP (1986) Profils neurologiques et mentaux dans les lésions encéphaliques diffuses post-traumatiques sévères, 103 cas. In Neurochirurgie 32: 63–73
17. Spitzer RL (1984) Psychiatric diagnosis. Are clinicians still necessary? In: Williams JBW, Spitzer RL (eds), Psychotherapy Research: Where are we and where should we go? Guilford, New York, pp 273–288
18. Sundberg ND, Tyler LE (1967) Clinical psychology: An introduction to research and practice. Appleton-Century-Crofts, New York
19. Webb EJ, Campbell DT, Schwartz RD, Sechrest L (1966) Unobtrusive measures: non reactive research in the social sciences. Rand McNally, Chicago
20. Weiner H (1978) The illusion of simplicity: the medical model revisited. Amer J Psychiat 135: 27–33
21. Wiggins JS (1973) Personality and prediction: principles of personality assessment. Addison-Wesley, Reading, Mass

Correspondence: Prof. Dr. W. Huber, Department of Psychology, University of Louvain, Louvain-la-Neuve, Belgium.

Summary

The object and the domain of the psychology of personality being less well defined and agreed upon than in other sciences, to discuss successfully the present situation of the methodology of the evaluation of the personality supposes the prior clarification of a certain number of questions.

This is done by conceiving of the psychology of personality as of an empirical science and by addressing what seems to be the four major relevant questions: 1) the major issues in the psychology of personality, 2) the major questions of the assessment of the individual personality, 3) the sources and methods of gathering data in this discipline, and 4) the process of clinical judgment.

Keywords: Personality; methodology; diagnosis; assessment; clinical judgment.

Acta Neurochirurgica, Suppl. 44, 54–55 (1988)

Discussion on Methods for the Evaluation of the Personality

Normal Brain/Damaged Brain

N. Sichez-Auclair

Centre du Langage, Service de Neurologie du Prof. F. Lhermitte, Hôpital de la Salpêtrière, Paris, France

Summary

The traditional way for the evaluation of the personality is not always adapted to a brain damaged population. The difficulties encountered by using standardized psychological tests and methods to remedy them are described.

Keywords: Psychological functioning; measurement techniques; control groups.

Introduction

Theorists and researchers in intercognitive psychology, computer science and brain theory work together to understand better how humans encode, represent and process sensory information and try to elaborate a suitable methodology. Likewise, with regard to the personality the fundamental psychological processing might be elicited for understanding the mental functioning.

The problem in its entirety stems from the fact to know whether a method worked out for a normal population is also adapted to a brain damaged population.

As the Glasgow Coma Scale (GCS) provides common nomenclature for measuring coma, a similar type of measure would be required to describe the psychiatric sequelae of brain damaged patients. Nevertheless, for them, tests such as Minnesota Multiphasic Personality, Inventory (M.M.P.I.) or Rorschach appear to have limited value, for they have generally been standardized on psychiatric patients. Consequently it remains unclear how patterns of test scores are affected by various forms of brain damage. Thus the elevation of "schizophrenia" on the M.M.P.I. may simply reflect confusion in thinking secondary to brain lesions. Moreover the validity of most of the tests or questionnaires is based on the patients capacity to be objective in self-reporting. Brain-injured patients often have little insight into their difficulties and using their psychiatric functioning may be misleading. To be sure, unrealistic self-appraisal is common after brain injury and is often overlooked by the traditional psychiatry or clinical psychological tests. Patients suffering frontal lobe damage are often unrealistic about the changes they have undergone.

Another source of errors stems from the fact that disorders of the cerebral functions may handicap the evaluation of the morbid personality. Of what use are the verbal personality tests in a patient with aphasic disorders? How does one interpret the data on the Rorschach Test in a patient showing a spatial hemi-attention. Is it a psychological denial? Moreover the need to be precise and to verify the genuiness of the reports led to questionnaires being filled out by relatives and by several psychologists, so that some problems may be met in this area. Thus, with regard to head injury, agreement between patients and relatives is generally high regarding sensory and motor impairments but it is very low on emotional and behavioural changes. According to the studies by Brooks it is much more likely that the patient will deny a problem which the relatives report than vice versa. Therefore disagreement would be not related to cognitive problems in the patients but to the relatives personality. Actually, under stress conditions, relatives would overestimate the effects of injury on the patient.

In the same way disagreement may be noted between physicians. It may arise for various reasons:

1) change in response of the patient when the examination takes place at different times;

2) observer bias such as errors in recording observations;

3) differences in eliciting the response;

4) differences in the interpretation of the same response (Maas *et al.*[1]).

It seems that the main bias is the time of assessment: thus, several works indicated that during the first six months after brain injury, relatives report a minimal amount of psychiatric disturbance; afterwards their reports often increase. In a recent study, Prigatano[3] reported the case of a few patients with a GCS Score of 13 to 15 on admission, who had evidence of psychiatric disturbances some months or years later.

Are There Any Ways to Remedy These Difficulties?

Among the methods possible, considering the neuropsychological context is an important one.

1) The patients behaviour during the neuropsychological assessment provides some clues to the possible neuropsychologically mediate disturbances.

For example, what role do different types of perceptual disturbances play in psychotic-like disorders after brain injury?

2) Another way is to constitute control groups. Why? For three main reasons: first of all because one methodological problem in determining the aetiology of psychological complaints is the absence of information on patient functioning before the onset of injury; secondly, the use of control groups may help disentangle specific and non-specific effects; thirdly, they may provide moderating variables for the behavioural problems of brain-injured patients.

Which Are the Types of Control Groups and How Can One Constitute them?

A control group differs from the experimental group with respect to the independent variable under study. Therefore it is necessary to obtain a group which is similar to the brain-damaged group in certain respects, yet has no brain damage. Thus concerning the traumatic population which is not a random one, the control group should firstly be drawn from a similar population at risk with overrepresentation of the young, of males, and of the lower socio-economic classes which are found in the brain-injured population. In this respect, young severe orthopaedic or paraplegic patients are representative. Pocok *et al.*[2] proposed a group consisting of the pretrauma friends. These subjects were used as potential control based on the assumption that one usually chooses friends similar to oneself "Birds of a feather flock together". Finally, reactionary problems may be outlined by using a group of subjects who have suffered catastrophic illness or who have undergone major losses in their lives or important changes in their job status.

In Conclusions

We need to construct a working model of personality disturbances after brain damage, whatever it is, and to develop appropriate measurement techniques to test components of the model. Research on psychiatric disturbances after brain damage should attempt to develop methods that would assess the reactionary neurologically mediated and characterological components.

Research in this fields clearly requires an interdisciplinary effort and careful classification of patients with both psychiatric and neurological disorders.

References

1. Maas AIR, Braakman R, Schouten HJA, Minderhoud JM, Zomeren AH van (1983) Agreement between physicians on assessment of outcome following severe head injury. J Neurosurg 58: 321–325
2. Pocok SJ, Simon R (1975) Sequential treatment assignment balancing for prognostic factors in the controlled clinical trial. Biometrics 31: 103–115
3. Prigatano GP, Pepping M (1987) Neuropsychological status before and after mild head injury. BNI Quarterly 3: 18–21

Correspondence: Dr. Nicole Sichez-Auclair, Centre du Langage, Service de Neurologie du Prof. F. Lhermitte, Hôpital de la Salpêtrière, 47 Boulevard de l'Hôpital, F-75651 Paris Cédex 13, France.

IV. Changes of Personality as Consequence of Severe Brain Injuries

Acta Neurochirurgica, Suppl. 44, 59–64 (1988)

Personality Change After Severe Head Injury

N. Brooks

University of Glasgow, Department of Psychological Medicine, Glasgow, U.K.

Summary

Personality change is widely reported after head injury, but rarely investigated quantitatively. This Paper summarises recent quantitative studies concerned with the natural history of personality change after severe head injury; its nature; its prediction; and its consequences. Reports of personality change increase with increasing time after injury, and the changes reflect a variety of phenomena including changes in affect, behaviour, maturity, and responsibility. Although there is a relationship between injury severity and personality change, other features such as premorbid personality and lifestyle are important. A severe personality change has important functional consequences. Families become heavily burdened, and patients with severe personality change are very unlikely to return to work.

Keywords: Personality; head injury; family; work.

Introduction

Reports of personality change after severe head injury are made frequently by both the injured patients themselves and their family members. This is not a static phenomenon, however. In a study in 1983 (Brooks and McKinlay 1983) we asked a close relative of 55 severely head injured patients to report the presence of pesonality change, in response to the simple question "Is he the same person after the injury as he was before?". By 3 months after injury, 49% were reporting a change, and this had increased to 60% by 6 and 12 months. In a 5 year follow-up of 43 of the original 55 patients the proportion had increased to 74%. This is a very high percentage indeed, but this is rather what clinical experience may lead one to expect, in these very severely injured patients. In each case their injury had been severe enough to warrant transfer from an initial receiving hospital to a regional neurosurgical unit. All of them had been in coma (defined in terms of the Glasgow Coma Scale) for at least 6 hours, or had had a post-traumatic amnesia of at least 48 hours, or had had an intracranial haematoma removed. All the patients to be described in this chapter meet these criteria.

The figures in the previous paragraph refer to reports by a relative of the injured patient. The reports from a relative and a patient inevitably differ, as may be expected, and a recent study (McKinlay *et al.* 1984), has shown that whereas patients and relatives differ rather little in the reporting of sensory motor changes, they do differ in their reporting of behavioural and personality change. Typically, patients when compared with relatives under-report personality change.

The rest of the chapter will be devoted to an analysis of the nature of the personality change, its prediction, and its functional consequences in terms including family distress and return to work. Throughout the chapter, the term "personality change", will be used when dealing with the general aspects of such a change, or with responses made by patients or family members to a specific question concerning the presence or absence of an overall personality change. The term "emotional/behavioural change", will be used to refer to some of the specific features which go to make up the general personality change.

The Nature of the Change in Personality

While the psychological literature contains many theories of personality, they often seem rather difficult to apply in the study of head injury, bearing little relationship to the clinical realities of post head injury disability. Where theories have led to specific personality measures, they all too often seem clinically inappropriate, or clinically unwieldy because of length and/or intrusiveness, or the need for highly skilled and occasionally idiosyncratic interpretation. The approach described in this chapter is a very pragmatic one in which attempts are made to identify the underlying elements of personality change by asking simple questions such as "what *exactly* had changed?", and "how does this change over time?". To do this, two separate approaches will be described. The first attempts to identify specific features (*e.g.* irritability, depression, etc.) which reflect the personality change in the patient. The second approach is to use a model derived essentially from semantic differential methodology, but tailored specifically for head injury. In this approach both patients and relatives (only the reports of relatives will be dealt with in this chapter) are invited to define the patient's "current", and pre-injury personality profile in terms of a series of bipolar adjectives of the type shown in Fig. 1.

Results from the first approach have by now been extensively reported by the Glasgow Group (McKinley *et al.* 1981, Brooks 1984, Brooks *et al.* 1987 a, 1987 b), and this section therefore will briefly summarize these results. Reports from relatives of severely injured patients very frequently refer to change in personality. When relatives are interviewed and asked to report the presence or absence of a wide variety of changes in the patient, they show themselves to be aware of changes even by 3 months after injury. In our studies, we have interviewed relatives at 3, 6, and 12 months, and asked them to report problems or features of behaviour which are now present, and which were not present before the injury. The changes we have tried to identify have included physical and cognitive (not to be dealt with here); subjective (symptoms such as tiredness, headache, poor tolerance of noise, etc.); dependency (both self care difficulties, and the need for supervision); and disturbed behaviour (difficult bizarre, puzzling, or inappropriate behaviours). Recent studies (Brooks *et al.* 1987 a, 1987 b) have indicated the relative frequency of change in these broad categories within 7 years of injury. One study involved a serial follow-up of 42 patients seen 3 times during the first year, and again at 5 years, and the other a study of 134 patients seen at any time between 2 and 7 years after injury. The figures in Table 1 refer to the former study, and indicate the "top 5" problems within 5 years of injury. These problems are nearly all behavioural/emotional or cognitive. Reports of personality change are frequent even at one year after injury, but by 5 years they represent the single most frequently reported change in the patient. Close to that frequency are difficulties in emotional inhibition (irritability and temper). The next 5 items in terms of frequency (all reported by at least 54% of relatives at 5 years) are as follows (the first figure is the first year, and the second the fifth year percentage); tiredness: 69%, 62%; depression: 51%, 57%; rapid mood changes: 57%, 57%; tension and anxiety: 57%, 57%; threats of violence: 15%, 54%. The very major increase in threats of violence against the relative is obvious and disturbing.

Table 1. *"Top 5" Problems Reported by Relatives of 42 Severely Head Injured Patients*

Problem	One year	Five years
Personality change	60%	74%
Slowness	65%	67%
Memory Disturbance	67%	67%
Irritability	67%	64%
Tempers	64%	64%

The picture in terms of broad categories of behavioural/emotional change is much as might be expected. By one year, the most obvious changes reported by relatives are broadly emotional and subjective, or cognitive. These and the other broad categories of outcome referred to above (emotional, dependency, etc.) were each scaled to give a maximum of 10, and the mean reports of emotional, and subjective change made by 42 relatives were 5.0 and 4.3 respectively at one year, and 5.2 and 4.2 at 5 years. These may be contrasted with the figures for memory disturbance of 1.6 and 2.9; and physical disturbance of 2.3 and 2.6. The other most frequently reported category of change at 5 years was disturbed behaviour, with a mean score of 3.7, showing a large increase from the one year figure of 2.2.

There is therefore no doubt that the reports of personality change per se, and of the specific features underlying change, occur frequently at one year, and become even more frequent at 5 years. The possible reasons for such an increase will be considered shortly. Before that, the other approach to the study of personality change will be discussed—the use of bipolar adjectives.

This methodology was reported by the author (Brooks and McKinlay 1983), and by Tyerman and Humphrey (1984). The Brooks and McKinlay report described the results of ratings of personality made by a relative of the patient, whereas Tyerman and Humphrey reported the ratings of patients themselves. Relatives are well able to understand this approach, and readily fill in the adjectives. The defining features of personality change can be assessed by identifying those relatives who do and those who do not report personality change in the injured patient, and then comparing their ratings on the adjectives are shown in Fig. 1. In

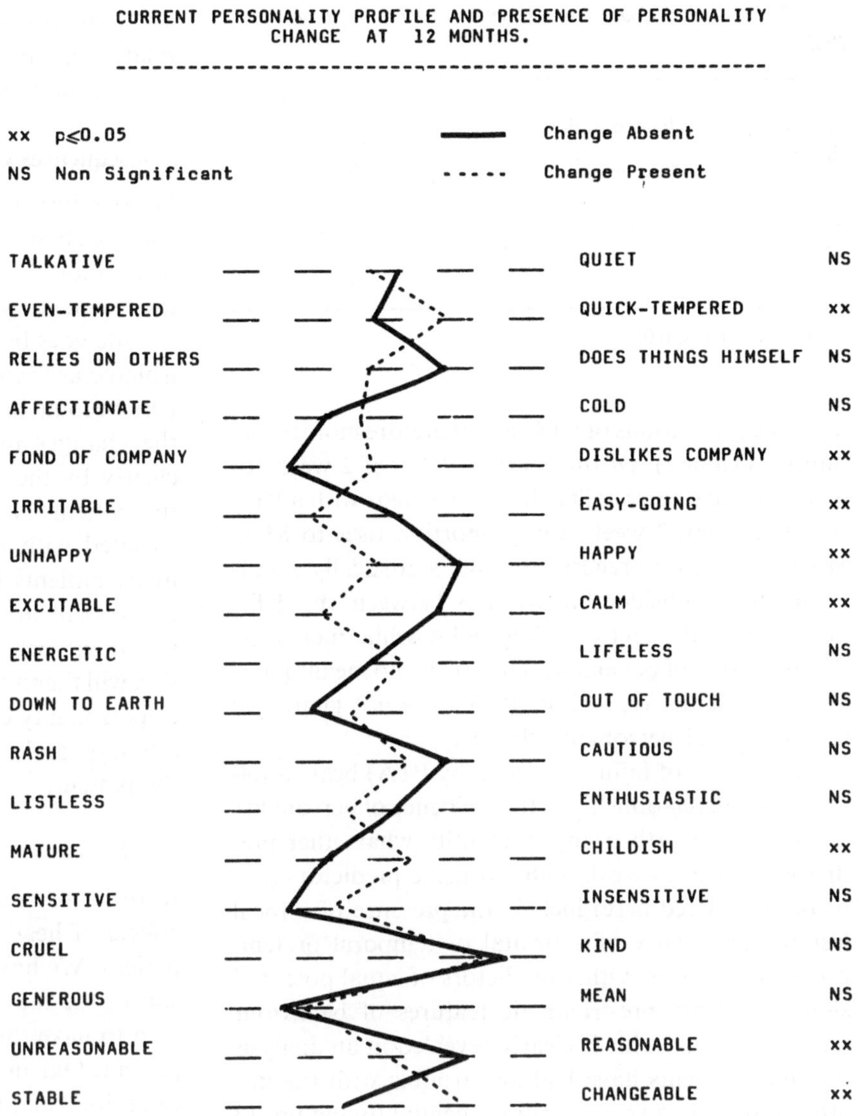

Fig. 1. Reproduced from a paper "Personality and behavioural change after severe blunt head injury—a relative's view" which appeared in the Journal of Neurology, Neurosurgery, and Psychiatry (1983) 46: 336–344, with the permission of the publishers

this figure the data refer to reports by the relative of the patient as he is "now" at 12 months after injury, for the two subgroups of patients with and without personality change. As may be seen from the figure, patients who are reported to have changed have become significantly ($p < 0.01$ on t-test) more quick tempered, unsociable, childish, etc. The changes are all in a broadly negative direction, with no evidence of any "improvement" in personality following head injury. However, clinical experience does show that in a very small proportion of cases (perhaps 5% or fewer), the head injury may lead to an improvement in personality, as a previously heavy drinking and aggressive man becomes quiet and passive, perhaps because the injury has resulted functionally in a prefrontal leucotomy.

Prediction of Personality Change

The prediction of who will and who will not change in personality is very difficult. It is equally difficult to predict which patients will be described as showing an increasing incidence of personality change as time goes by.

One obvious possible predictor change is the severity of injury. This may be assessed in a number of ways, but one simple way is to use the duration of post-traumatic amnesia (PTA), which relates to a variety of aspects of outcome (Brooks, 1984). When patients with and without personality change (seen from 2 to 7 years after injury) are compared in terms of PTA duration, it can be seen that those with change are likely to have

Table 2. *Personality Change as a Function of PTA Duration in 122 Patients*

Personality change	PTA duration		
	Two weeks or less	Two–four weeks	Over four weeks
Absent	14	7	8
Present	20	57	16

$x^2 = 12.71$ $p < 0.01$.

had longer durations of PTA, and therefore more severe injuries (Table 2). Of those with a PTA of 2 weeks or less 59% have personality change, whereas with a PTA of longer than 2 weeks the proportion rises to 83%. However, the differences although statistically significant show considerable overlap between the PTA groups, and the fact of a long PTA while increasing the likelihood of personality change, by no means guarantees it. Similarly, a short PTA does not guarantee the absence of personality change.

As severity of injury (assessed by PTA) bears a relatively weak relationship to the incidence of personality change, it is worth trying to identify what other predictors may be at work. Other organic predictors (not to be considered here) include the presence of a focal lesion, particularly of a frontal or temporal or temporolimbic nature. Other predictors of equal potential significance are pre-traumatic features of behaviour and personality. This is clearly revealed in an analysis of which patients have had any trouble with the law after injury. By 5 years in our sequential follow-up, 13 (31%) of our 42 patients (7% at one year) had had some trouble with the law, with the problems ranging from breach of the peace to attempted murder. It is tempting to attribute these offences entirely to the effects of the injury, but in the case of 9 patients there had already been trouble of some nature even before the injury, giving at least a 21% pre-injury conviction rate. These figures also raise the phenomenon of attribution, in which any negative post-traumatic feature is attributed to brain damage caused in the injury. Attribution is a very pervasive process, as both patient and relative struggle with the existentional crisis caused by severe brain injury, and attempt to impart meaning to the injury and its consequences. Other mechanisms leading to reports of personality change (particularly increasing reports as time goes by) include secondary reactive changes in the patient as he contemplates the life he has now compared with the life before injury, and reduction in denial (Romano 1974) by patient and relative as an unpleasant reality becomes increasing evident, and increasingly less able to be ignored.

Consequences of Personality Change

Head injury triggers a large number of changes, including changes in social and family interaction. Many head injured patients report that they no longer have the friends they had before, and this problem increases as time goes by (Kozloff 1987). Families often become isolated as friends stop visiting, and many family members become increasingly distressed. This distress, and the changes in social family life are triggered very clearly by the emotional and behavioural changes in the patient. Not only is social and family distress associated with personality change in the patient, but many patients fail to return to work after injury, and as we shall show below, this failure is related to emotional and behavioural change in the patient. This section will therefore consider the functional consequences of personality change firstly in terms of distress within a family, and secondly, in terms of return to work for the patient.

(a) Family Distress

In order to address this area we have considered the effects of head injury upon the family as well as the patient. We have concentrated particularly on the person in the family who, immediately after injury, was seen to have the main responsibility for caring for the patient. Our initial study concerned 55 patients and a close informant followed up for one year, with the informants described in Table 3.

We used a model in which changes after injury are designated "objective burden" (objectively observable changes in the patient, or in family interaction, etc.), and "subjective burden" (distress, tension, etc. in family members). A major concern has been to identify the causal links between objective and subjective burden. Subjective burden has been assessed in a number of ways, and one very simple approach has been to ask the relative to fill in a seven point analogue scale rang-

Table 3. *Relationship to Patient of Informant at One Year*

Wife/husband	N = 31
Mother/father	17
Other relative	5
Not a relative	2
	55

ing from a score of one (no stress) to seven (severe stress) in which the relative rates how he/she is feeling as a result of coping with the consequences of the injury in the patient. Relatives can then be categorized as low subjective burden (SB 1–2); medium (3–4); or high (5–7), and the distribution and predictors of such burden can then be identified. At one year, most relatives fall in the moderate or low burden category (Brooks *et al.* 1986); with 43% low; 33% medium; and 24% high. At 5 years, the picture is very different, with only 10% in the low category, 33% in the medium, and 56% in the high subjective burden category.

The best predictor of subjective burden in the relative at 1 and 5 years after injury is the presence of a severe behavioural/emotional change in the patient. For example, at one year, there were significant relationships between the magnitude of burden in the relative, and the level of emotional, subjective, behavioural, and physical and cognitive change in the patient. At 5 years, the picture was very similar except that physical changes no longer bore any relationship to burden in the family member, whereas dependency changes did. More recent studies of a group of 134 severely injured patients seen from 2–7 years after injury (Brooks *et al.* 1987) have extended this finding, by identifying very specific changes in the patients which relate to high burden in the family member. These results show that a wide range of negative emotional and behavioural changes in the patients are related to high subjective burden in the relative, but the relationship is by no means symmetrical as can be seen in Table 4. This Table shows the relationship between subjective burden in the relative and dependency change in the patients. The presence of high dependency in the patient is a good indicator of high burden in the relative. The absence of the dependency deficit in the patient does not guarantee that there will be low burden in the relative. Indeed, the relative may still be experiencing substantial degrees of burden.

The relationship between subjective burden and personality change can also be addressed using the bipolar adjective methodology reported earlier. To do this (Brooks and McKinlay 1983) relatives were subdivided into those who reported mild—medium burden at 1 year (N = 40), and those reporting high burden (N = 15). The adjective scores of patients associated with the two groups of relatives were then compared using t-test. Many of the comparisons were significant, showing that a high degree of subjective burden in the relative was associated with changes in the patient which included increased temper, social withdrawal, emotional coldness, lack of energy, and increased cruelty, meanness, unreasonableness, immaturity, intensitivity, and changeability.

Table 4. *Subjective Burden in a Relative as a Function of the Level of Dependency Deficit in the Patient*

Dependency deficit	Subjective burden		
	Low	Medium	High

(b) Vocational Outcome

Return to work is a doubtful index of general outcome, because of the widely differing patterns of unemployment both within and between different countries, and because of the many possible reasons for a failure to return to work. Nevertheless, the simple facts speak for themselves in that in our study of 134 patients seen between 2 and 7 years after injury, 86% of patients were employed before the injury, and only 29% after. A job gives a sense of dignity, worth, and esteem within a community, as well as the obvious financial rewards. A severely injured patient has to face not only the consequences of the injury, but also the secondary consequences of loss of work.

In our recent study, demographic features were of obvious importance in predicting return to work (age, employment status before injury), but the presence of behavioural/emotional deficits (as well as cognitive deficits) was a very significant predictor to failure to return to work, as has been reported by others (Weddell, Oddy, and Jenkins 1980). In our study, two features of personality proved to be particularly significant. The first was the presence of a dependency deficit, and the second the presence of an emotional deficit. The first deficit may not seem initially like a personality feature. However, detailed study showed that there were specific features about the patient which led a relative to consider that he was no longer as independent as before injury. The features that concerned the relative were that the patient was no longer able to take responsibility, or had problems with personal hygiene. The latter appeared to result from a personality rather than physical change; that is the patient cared less about personal presentation, rather than having a physical inability to wash or dress appropriately.

The emotional deficit of particular significance in predicting failure to return to work was a difficulty in the control of anger. Just as in the prediction of subjective burden, the relationships between features in the patient and return to work were not symmetrical. That is, the presence of a negative change in the patient was a good predictor of a failure to return to work. The absence of a negative change was no guarantee of return to work.

Summary and Conclusions

Personality changes as a consequence of severe head injury are common. They may increase over time if not treated, and have very real negative social and family consequences. The prediction of exactly which patient will show which change is far from easy, as is the prediction of the precise negative consequences of change in personality. However, in the case of the latter pediction, asymmetric relationships seem to be the rule, with the presence of a negative change in the patient signalling a very high likelihood of negative consequences, but the absence of such a change not conferring immunity from negative consequences.

Studies which have identified the nature, evolution, and consequences of personality change in the patient form the basis for strategies of rehabilitation intervention with severely injured patients. A rehabilitation regime which concentrates on behavioural/emotional changes from very early after injury may be very effective indeed in reducing the very serious long-term morbidity currently found in both patient and family in the absence of appropriate rehabilitation.

References

1. Brooks N (ed)(1984) Closed head injury: psychological, social, and family consequences. OUP, Oxford
2. Brooks N, McKinlay W (1983) Personality and behavioural change after severe blunt head injury—a relative's view. J Neurol Neurosurg Psychiatry 46: 336–344
3. Brooks N, Campsie L, Symington C, Beattle A, McKinlay W (1984) The five year outcome of severe blunt head injury: a relative's view. J Neurol Neurosurg Psychiatry 49: 764–770
4. Brooks N, Campsie L, Symington C, Beattle A, McKinlay W (1987) The effects of severe head injury on patient and relative within seven years of injury. J Head Trauma Rehab 2 (3): 1–13
5. Brooks N, McKinlay W, Symington C, Beattle A, Campsie L (1987) Return to work within the first seven years of severe head injury. Brain Injury 1: 5–19
6. Kozloff R (1987) Networks of social support and the outcome from severe head injury. J Head Trauma Rehab 2 (3): 14–23
7. McKinlay W, Brooks N, Bond MR, Martinage D, Marshall MM (1981) The short term outcome of severe blunt head injury as reported by relatives of the injured person. J Neurol Neurosurg Psychiatry 44 (6): 527–533
8. McKinlay W, Brooks N (1984) Methodological problems in assessing psychosocial recovery following severe head injury. J Clin Neuropsychol 6 (1): 87–99
9. Romano MD (1974) Family response to traumatic head injury. Scand J Rehab Med 6: 1–4
10. Tyerman A, Humphrey M (1984) Changes in self concept following severe head injury. Int J Rehab Research 7 (1): 11–23
11. Weddell R, Oddy M, Jenkins D (1980) Social adjustment after rehabilitation: a two year follow-up of patients with severe head injury. Psych Med 10: 257–263

Correspondence: Prof. N. Brooks, Ph.D., University of Glasgow, Department of Psychological Medicine, 6 Whittinghame Gardens, Great Western Road, Glasgow, 612 OAA, U.K.

Acta Neurochirurgica, Suppl. 44, 65–66 (1988)

Post-Traumatic Personality: as Many Cases as Individuals

N. Sichez-Auclair

Centre du Langage, Service de Neurologie du Prof. F. Lhermitte, Hôpital de la Salpêtrière, Paris, France

Summary

Post traumatic personality do not present a unique picture because the disturbances have many causes. That is why models structured basically to the understanding of the disorders must be carried out.

Keywords: Personality; basic mechanisms; aetiological factors.

Post-traumatic psychological symptoms are particularly complex for they do not present a unique picture. Thus, amongst brain-damaged patients the variation in behaviour tends to be much greater than that found within the non brain-damaged population. This heterogeneity is likely to be due to the fact that the disturbances have many causes. Thus, the potential aetiological factors contributing to psychiatric disorders after traumatic head injury listed by Lishman[5] showed this complexity very well. In other terms there are as many cases as individuals. That is why difficulties are encountered when attempts are made to relate clinical assessment of personality to a unitary concept of brain damage. In the literature of clinical psychology many studies have appeared which attempt to demonstrate the correlation of a particular personality characteristic or group of traits with brain damage.

Another source of error stems from the use of psychiatric terminology. Obviously head injured are not a random population; they have distinct personal and social characteristics which seem to predispose them to injury, although of course only very few of them were admitted to psychiatric departments before the injury. So, whatever the mechanisms, the psychological disturbances are related to the existence of the head injury.

For example, if the post-concussional syndrome poses some problems it is because it appears not to be related to the organic pathology. Its symptoms do not appear to be associated with cerebral lesions. They look like common psychiatric disturbances and particularly neurotic ones. Similarly disorders which are like classic psychiatric syndromes are seen in patients suffering mild or no cognitive impairment. Since an inverse relationship between severity of injury and severity of emotional symptoms has been demonstrated by several authors, whereas other authors found a positive correlation[1, 4], attention must be paid to neurotic-like disorders such as depression and anxiety. Besides, these kinds of psychological disorder are usually seen in traditional psychiatry; the other post-traumatic ones are more specific.

Thus, head-injured patients are not psychiatrically ill yet despite this fact, psychiatric terminology is used to describe their disturbances. The psychiatric terms refer to a defined illness but one may not speak about homogeneous lesions with regard to the post-traumatic psychological disorders. Therefore, can paranoia associated with post-traumatic temporal lobe damage be compared to the psychological structure of paranoia? Is the association of post-traumatic disorganized thinking and social withdrawal likely to be psychiatrically mediated? Moreover one of the consequences of the post-traumatic amnesia is the inability to construct an intelligible whole out of a fragmented experience. Do the illusions, hallucinations and ideational distorsions have a psychiatric origin? It is easy to see how feelings of suspicion become projected to the environment leading to the development of a paranoid state.

Likewise too rigid attempts to link the psychiatric disturbance to specific anatomical areas may be another source of error, for, traumatic damage is often so widespread that it reaches several brain areas and in addition the same disorders may be encountered in connection with different types of brain lesion. This reductionist attitude, may be very unhelpful.

Changing post-traumatic nomenclature enables one to develop a model structured basically to the understanding of the disorders. So, let us observe the behaviour of severely brain injured. In the cases of immediately appearing coma due to impairment of the brain by abrupt acceleration and deceleration of the head, the only constant psychological dimension is a global reduction or a global disinhibition. Perhaps these terms refer to the dimension introversion-extroversion? The reduction of speed of information processing is often associated with a trend to mental inertia. This reduction is likely to be consistent with the neuroanatomical reports of axonal degeneration in monkeys that have undergone concussive injuries. Disorders of arousal are important for producing decreased motivation. This may help explain that a hallmark of brain injury is a less sophisticated emotional motivational being. Given this terminology, it is conceivable that different structures might exist in the brain stem and midbrain which serve the function of mediating the features of personality.

The Popper and Eccles conceptions which are rich in neurobiological and philosophical arguments may yield a useful model for understanding how a severe head injury may lead to a personality change. According to this theory the self-conscious mind and its brain are partners of a permanent dialogue; the personality expresses the continuity of the neuronal network.

The classification of psychological disorders according to the basic mechanisms might permit a more complete and reliable understanding. Thus: an impairment of mental control may explain the existence of some disturbances. Such psychological traits, which are not consistent with a normal social life, are likely to appear after an injury. A defect of control is characterized by the absence of inhibition occurring with a variety of affective and behavioural patterns. Consequently, the control over sudden shifts in mood and basic drives is abolished. Moreover the psychoaffective regression, partly characterized by childishness, may be explained by the impairment of developmentally more recent and more fragile structures. This hypothesis suggests the same phenomenon as the neurological recovery associated with the caudorostral restructuration ... This deep regression is supposed to express the life of the self-conscious mind in scanning an actual cortical activity without connection to the coded informations of the past. Likewise the self-consciousness is disturbed when certain lesions impair the access networks to references necessary for the memory. This may be the case in psychotic-like states.

A new approach involves the neurochemical abnormalities following traumatic brain injury. So certain depressive disorders in traumatically injured patients may be related to neurotransmitter disturbances. Some studies prove this to be a promising area of investigation[2, 3, 7]. Studies of adrenal cortex response to craniocerebral trauma have suggested elevated cortisol secretions. Serum cortisol abnormalities frequently observed during the first few weeks after trauma are also related to the degree of intracranial pressure (Steinbock and Thompson[6]); elevated serum cortisol has been reported in patients who experience hightened stress and also in endogeneously depressed patients. A fruitful area of study could be one concerning the relationship of various metabolic and endocrinological changes associated with craniocerebral personality and behavioural disorders.

In conclusion, it is not always clear if psychological impairment is a reaction to the neurologic or neuropsychological deficits or if it reflects a preexisting personality disorder or altered chemistry of the brain. How do the commonly incurred orbital frontal and anterior temporal lobe injuries suffered in acceleration-deceleration closed head injury interact with sociopathic personality characteristics to produce the picture of symptoms observed several months after the injury?

The quality of the answer depends on the quality of the methods employed and need to renounce the classical dualism which opposes mind to matter.

References

1. Brooks N (1984) Closed head injury. Psychological, social and family consequences. OUP
2. Clifton GL, Robertson CS, Grossman RG *et al* (1984) The metabolic to severe head injury. J Neurosurg 60: 687–696
3. Deutschmann CS, Konstantinides FN, Raup S, Thienprasit P, Cerra FB (1986) Physiological and metabolic response to isolated closed-head injury. Part 1: Basal metabolic state: correlations of metabolic and physiological parameters with fasting and stressed controls. J Neurosurg 64: 89–98
4. Kinlay MCWW, Brooks DN, Bond MR (1983) Postconcussional symptoms, financial compensation and outcome of severe blunt head injury. J Neurol Neurosurg Psychiatry 46: 1084–1091
5. Lishman WA (1978) Organic psychiatry. The psychological consequences of cerebral disorders (Aetiological factors in psychiatric disturbance after head injury). Blackwell Scientific Publications, Boston
6. Steinbock P, Thomas G (1979) Serum cortisol abnormalities after cranio-cerebral trauma. Neurosurgery 5: 559–565
7. Uzzell B, Brist WD, Dolinskas CA, Langfitt ThW (1986) Relationship of acute CSF and ICP findings to neuropsychological outcome in severe head injury. J Neurosurg 65: 630–635

Correspondence: Dr. Nicole Sichez-Auclair, Centre du Langage, Service de Neurologie du Prof. F. Lhermitte, Hôpital de la Salpêtrière, 47 Boulevard de l'Hôpital, F-75651 Paris Cédex 13, France.

Acta Neurochirurgica, Suppl. 44, 67–69 (1988)

Post-Traumatic Psychoses

A. Violon

Department of Neurology, Hôpital Saint-Pierre, Brussels, Belgium

Summary

The incidence of post-traumatic psychoses can be appraised as 3–4 per cent in the adults who suffered a head injury.

Post-traumatic psychoses appear as not directly provoked by the head injury but rather as precipitated by it, usually in young male patients with previous psychological disturbances which were more or less compensated formerly.

In half the cases, the psychosis described as an acute delusional state totally disappears after a few days or weeks.

Keywords: Head injury; personality; psychosis.

Post-traumatic psychoses have for a long time been a controversial matter, noteworthy because many authors have gathered under the heading of psychosis different types of severe post-traumatic mental disturbances such as confusional states[9], dementia or Korsakoff's syndrome[1, 2, 8].

The lack of an accurate definition of post-traumatic psychoses has made the comparison of the data in this field harzardous but explains why the findings in the literature appear so variable.

In order to dispel the controversies and misunderstandings surrounding this concept, Brihaye *et al.* proposed, in 1979[4], the following definition:

"Post-traumatic psychoses are regressive or chronic acquired delusional states appearing after a head injury, in non-demented patients, independently of the presence or absence of neurological disorders".

Three questions can be posed about post-traumatic psychoses.

I. Do Post-Traumatic Psychoses Really Exist?

In the popular fantasy, there is no doubt that a person can become "mad" after his or her head has been violently knocked. Indeed, post-traumatic psychoses do exist (1 to 12) but their incidence is very rare (Table 1). Using the rather restrictive definition already mentioned which excludes dementia as well as confusional states, we found that about 3,5% of post-traumatic patients became psychotic[4, 6, 12]. The psychosis does not in all cases start immediately after the accident or the awakening from coma (Table 2): There may be a delay, even a rather long one.

Table 1. *Incidence of Post-Traumatic Psychoses*

%	Number of cases of head injury	Authors	Year
0.07	5,798	Moros	1944
2.2	1,829	Hillbom	1951
8.9	3,552	Achte, Hillbom and Aalberg	1967
4.2	10,000	Vigouroux et al.	1972
< 1	244	Assal and Muller	1973
3.7	427	Brihaye, Violon and De Mol	1982
3.4	530	De Mol, Violon and Brihaye	1987

Table 2. *Interval Between the Accident and the Onset of the Psychosis*

Interval	Acute delusion	Chronic psychoses	Total	%
Immediately	4	6	10	62
Less than 2 months	2	1	3	19
2 to 23 months	0	0	0	0
24 to 48 months	2	1	3	19

From Brihaye, Violon and De Mol 1982.

II. What Are the Characteristics of the Post-Traumatic Psychoses?

Post-traumatic psychoses can consist of acute, reversible episodes. or chronic, long-lasting conditions (Table 3). In our experience, a duration of five months was the longest found for these acute states after which

Table 3. *Evolution of the Psychoses*

Duration of the acute delusion	less than 2 weeks	from 2 weeks to 5 months
N = 9	4	5
Evolution of the chronic psychoses	no progress	slight recovery
N = 9	6	3

From De Mol, Violon and Brihaye 1987.

Table 4. *Types of Psychotic Decompensations*

	Acute delusion	Chronic psychoses	Total	%
Paranoid disorders	1	5	6	33
Schizophrenic disorders	5	1	6	33
Manic episode	1	2	3	17
Brief reactive psychosis	2	0	2	11
Major depressive episode	0	1	1	6

From De Mol, Violon and Brihaye 1987.

Table 5. *Characteristic Symptoms of the Psychoses*

Disturbances	Acute delusion	Chronic psychosis	Total
Delusions of persecution	3	6	9
Manic behaviour	3	3	6
Visual hallucinations	3	3	6
Autism	4	1	5
Hetero- or auto-aggression	2	3	5
Depression	1	4	5
Agitation	2	2	4
Sexual disturbances	2	2	4
Onirism	2	1	3
Megalomania	1	1	2
Auditory hallucinations	2	—	2
Delusion of pregnancy	1	—	1

From Brihaye, Violon and De Mol 1982.

the delusion totally disappeared. In longer durations, the psychotic condition remained[4, 5, 6, 12].

As to the types of psychotic decompensations, schizophrenic disorders are more frequent in acute delusional states and paranoid disorders predominate in chronic psychoses (Table 4).

The most frequently observed symptoms are persecutory delusion and severe thymic disorders, followed by hallucinations, autism and also hetero- and auto-aggression (Table 5).

3. What Are the Determinants of Post-Traumatic Psychoses?

The most difficult question to answer concerns what makes a person become psychotic after a head injury. The authors opinions about this are variable, ascribing an importance to the severity of the brain lesion and the damage to the temporal lobes[1, 2, 7], the cerebral oedema[9] or the previously disturbed personality[1, 4, 5, 6, 10, 12].

About the site of the injury, our studies[4, 6, 12] confirmed the damage to the temporal lobe in 31 to 39% of the post-traumatic psychotic patients. Temporal lobe damage however is known to be frequent in traumatic brain injuries, so that this feature could be unspecific.

As to the severity of the head injury, common sense would assume that it ought to play a role in the emergence of a post-traumatic psychosis. However, after having carefully checked our cases, we could not demonstrate that severity played any relevant role (Table 6), and moreover it was irrelevant to the ensuing clinical course.

What about other possible determinants? (Table 7.) Most post-traumatic psychotic patients are males and young. A very relevant factor consists of the existence of previous psychopathological disturbances which were found in 83% of the cases[12].

Table 6. *Severity of the Injury*

Duration of coma	Acute delusion	Chronic psychoses	Total	%
⩽ one day	5	6	11	61
2 to 5 days	2	2	4	22
6 to 17 days	0	0	0	0
18 days to 3 months	2	1	3	17

From De Mol, Violon and Brihaye 1987.

Table 7. *Possible Determinants of Post-Traumatic Psychoses*

Type of head injury	: irrelevant
Severity of head injury	: irrelevant
Site of head injury	: temporal in 39% of the cases
Previous psychopathological disturbances	: 83% of the cases
Post-traumatic psychotic patients are mostly	: younger than 30 (61%) : males (83%)

From De Mol, Violon and Brihaye 1987.

Conclusion

Post-traumatic psychoses are rare, not very well known, more frequent in young male patients, in the immediate period following the injury, commonly paranoid or schizophrenic and more frequent when the temporal lobe has suffered.

In our opinion however, post-traumatic psychosis ought rather to be called "post-traumatic psychotic decompensation". The accident indeed usually appears a trigger of pre-traumatic psychopathology.

References

1. Achte KA, Hillbom E, Aalberg V (1967) Post-traumatic psychoses following war brain injuries. A follow-up study on the psychoses developed by the men who suffered brain injuries in the Finnish War of 1939–1945. Reports from the rehabilitation Institute for Brain-Injured Veterans in Finland. 1: 1–101
2. Achte KA, Hillbom E, Aalberg V (1969) Psychoses following war brain injuries. Acta Psychiat Scand 45: 1–18
3. Assal G, Muller C (1973) Délires de longue durée lors d'affections traumatiques et tumorales du système nerveux central. Arch Suisses Neurol Neurochir Psychiat 112: 115–121
4. Brihaye J, Violon A, De Mol J (1982) Post-traumatic psychoses. Actas del XVIII Congresso Latino-Americano de Neurocirurgia. VI Congresso Latino-Americano de Neuroradiologia. Buenos-Aires 1979, pp 9–12
5. De Mol J, Violon A, Brihaye J (1982) Les décompensations schizophréniques post-traumatiques. A propos de 6 cas de schizophrénie traumatique. L'Encéphale 8: 17–24
6. De Mol J, Violon A, Brihaye J (1987) Post-traumatic psychoses: a retrospective study of 18 cases. Archivio di Psicologia Neurologia e Psichiatria 3: 336–350
7. Hillbom E (1951) Schizophrenia-like psychoses after brain trauma. Acta Psychiat Scand [Suppl] 60: 36–47
8. Joynt RJ, Shoulson I (1985) Dementia. In: Heilman KM, Valenstein E (eds) Clinical neuropsychology. 2nd ed. Oxford University Press, New York Oxford
9. Mifka P (1976) Post-traumatic psychiatric disturbances. Injuries of the brain and skull. Handbook of clinical neurology. Vinken and Bruyn Publ, Amsterdam Oxford New York, Part II, vol 24: 517–574
10. Moros N (1944) Traumatic psychosis: a questionnable disease entity. J Nerv Ment Dis 99: 45–55
11. Vigouroux PB, Baurand C, Choux M, Guillermain P, Chaix C, Deyts JP, Mignard P, Naquet R, Penciolelli R, Revillon J, Choux R, Mancia D (1972) Etat actuel des aspects séquellaires graves dans les traumatismes crâniens chez l'adulte. Neurochirurgie 18, [Suppl] 2: 254
12. Violon A, de Mol J (1987) Psychological sequelae after head traumas in adults. Acta Neurochir (Wien) 85: 96–102

Correspondence: Dr. A. Violon, Department of Neurology, Hôpital Saint-Pierre, Rue Haute, 322, B-1000 Brussels, Belgium.

Acta Neurochirurgica, Suppl. 44, 70–73 (1988)

Personality After Head Injury

R. A. Frowein and **R. Firsching**

Neurosurgical University Clinic, Köln, Federal Republic of Germany

Summary

For the discussion of post-traumatic so-called full recovery, only a very limited array of injured persons were at our disposal. Two-thirds of these patients are young and their personality needs long, patient rehabilitation. However, duration of coma in young severely injured patients does not seem to mean the same as in elderly patients.

Of the last third of our patients the greatest number work as craftsmen. Among the patients with a higher level of education, there were only 7% of patients with full recovery and also 7% with partial recovery, who actually went back to their original work.

For the assessment of results, we have to be aware of the important differences of age and level of education. Therefore it would be probably worth-while to continue the work of Jean Brihaye in the Glossary, to define the items of post-traumatic personality, so that for a strongly restricted assessment of results, a limited but clearly defined group of patients can be compared.

Definition

The preceding papers have defined changes in personality as part of the post-traumatic syndrome with sequelae beyond purely neurological disorders. The clinical psychologist Walter Poppelreuter in Bonn, 1915–1937, coined the term „Hirntraumatische Leistungsschwäche". Hans Wieck, Köln and Erlangen, 1977, stressed the fact, that in most of the cases the neuropsychiatric disorders are only *temporary*. Therefore, he called them a "transient syndrome", in contrast to the *permanent* state of neuropsychiatric deficit which is "dementia".

Factors Involved

Among the multiple factors which influence the sequence of events we would like to draw your attention to the degree and to the duration of coma, to the age of the patient and to his education and profession.

An important aspect for the evaluation of the post-

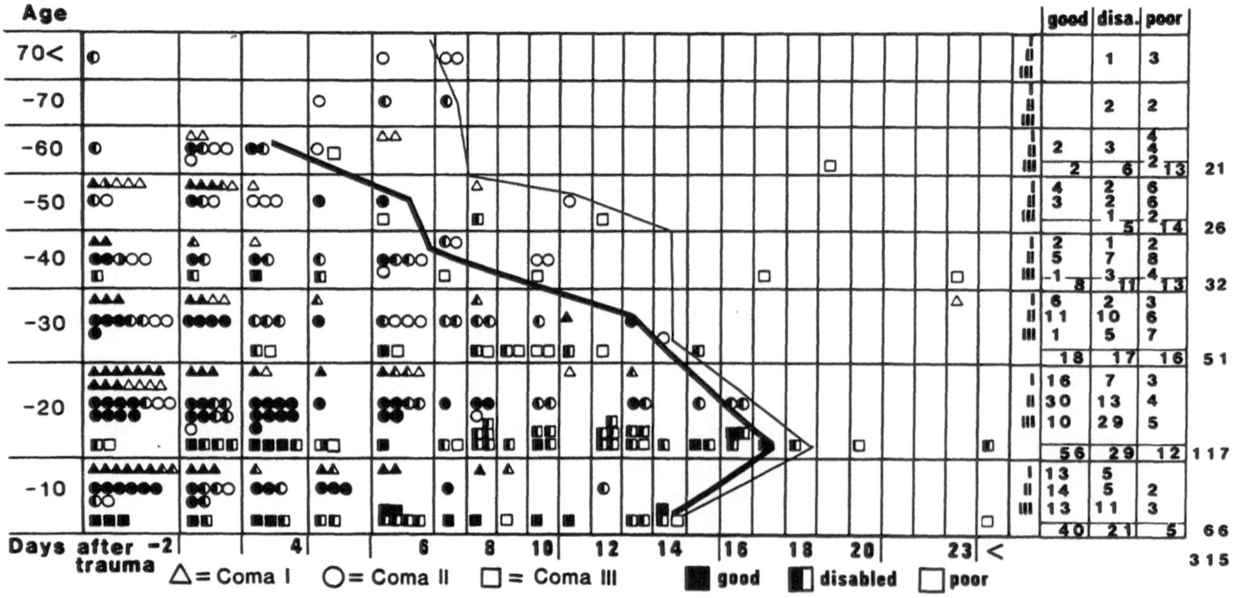

Fig. 1. Outcome of 315 survivors of long-lasting coma

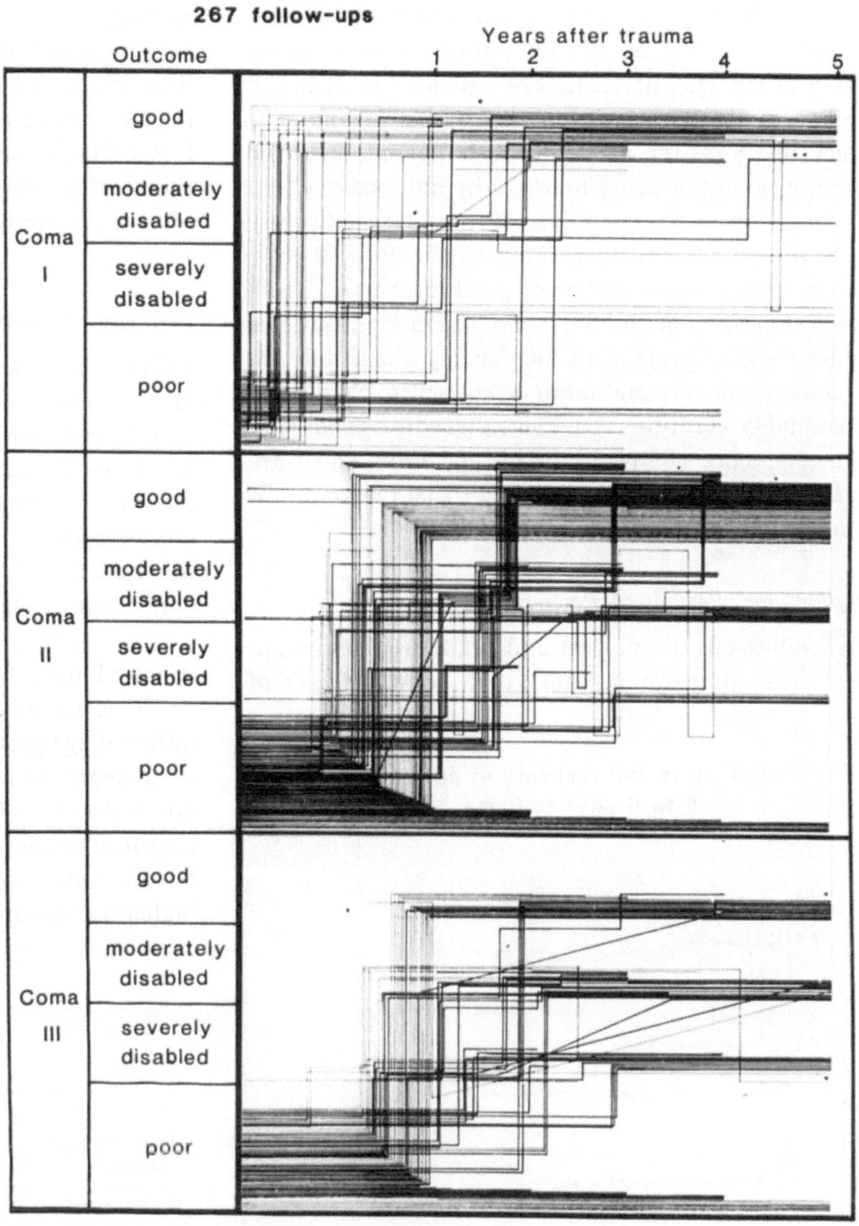

Fig. 2. Follow-up of 267 patients with long-lasting coma

traumatic personality is the recovery of the ability to work and good vocational resettlement.

Hypotheses

Based on the observations of 465 survivors of long-lasting post-traumatic coma our message is threefold:

— The chances of full recovery without severe changes in personality are limited by age and duration of coma.

— The period of full recovery of personality depends more on the degree of coma than on age.

— The number of patients with higher education and long-lasting coma is small.

Chance of Survival and of Good Recovery

In previous investigations we have already shown (Frowein *et al.* 1987) that after long-lasting coma the chance of survival and the chance to reach full recovery are limited by the age of the patient and the duration of coma.

The *5% survival limit*—thin line in Fig. 1—varies according to age from seven days of coma for 50 to

70-year-old patients in coma grades I and II to 18 days of coma for 10 to 20 year-old patients in coma grade III. Among 315 survivors, 124 patients (39%) had a full recovery; they are identified by black dots. The thick line joins the points of those patients with the longest period of coma followed by full recovery.

In young age groups this *borderline of good recovery* is near to the 5% survival limit, but in 40 to 60 year-old patients, the borderline of good recovery is 3 to 8 days shorter than the 5% survival limit. In patients older than 60 years with long-lasting coma only incomplete recovery and more or less severe changes of personality were observed. This result corresponds with the experience of Brooks (1984) and Edna and Cappelen (1987), who stated that bad social outcome was closely related to greater age.

Follow-ups: Periods of Recovery

267 follow-ups of one year and 178 longer follow-ups to 5 years after trauma show, that the final degree of

recovery—poor, moderately or severely disabled (Jennet and Bond, 1975)—was usually reached during the first year after trauma, in only 20% of the cases during the 2nd year or later. After post-traumatic coma grade I (GCS 6–7) the mean period of full recovery was shorter than after coma grade II (GCS 5) and especially after coma grade III (GCS 4) (Fig. 2).

Different Age—Same Duration of Coma

Taking the factor of age into consideration, there is also a rather unexpected result in our series:

Figure 3 shows the follow-up of 23 patients of different age at trauma, but all with the same duration of coma of 4 to 6 days.

The periods of time between trauma and full return to work, were 3 to 15 months in about all age groups with coma grade II (GCS 5).

After coma grade III (GCS 4 or less) the period was usually longer, from $1^1/_2$ years up to 5 years.

Thus, rather unexpectedly for us, for patients in different age groups who survived the same length and same degree of coma, the period of recovery was also about the same. Older patients who survived the same length of coma recovered as quickly as younger ones, but the number of these survivors is very small in the higher age groups.

Period up to full recovery of patients 4 to 6 days in Coma

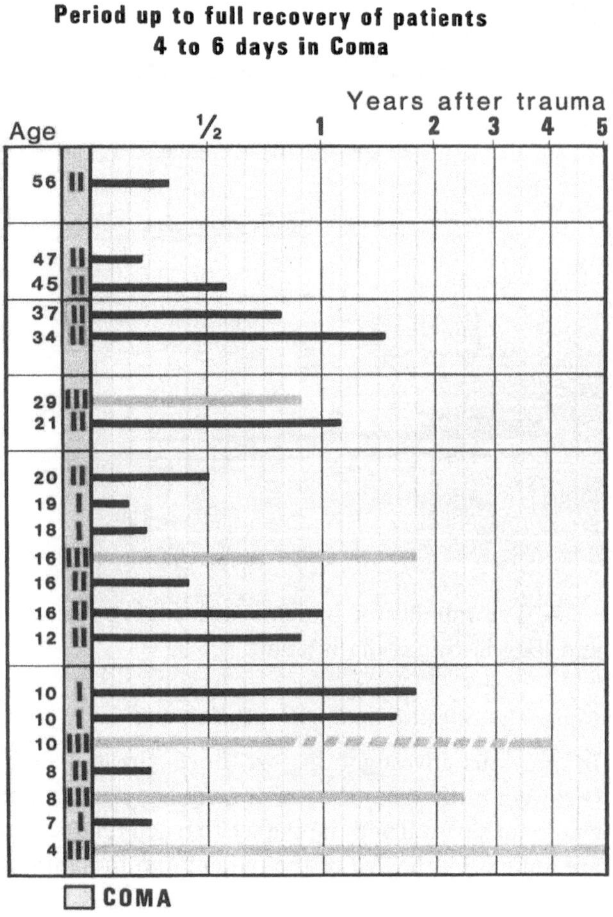

Fig. 3. Period up to full recovery of patients who were four to six days in coma

Profession	n	full recovery							partial
		Days of Coma							
		-3	-6	-9	-12	-15	-18		
Manager	1	□ 1							2
Businessman	6	□□ 2	□□□ O 4						5
Craftsman	9	□□□□ 4	□□□ 3	□ 1			□ 1		9
Workman	9	□□□ O △ 5	□ △△ 3		O 1				10
Housewife	4	□□ 2	□□ 2						4
Pensioner									2
Secretary	3	O 1		O 1		△ 1			
Student	4	OOO 3		O 1					3
Apprentice	16	△△△△△ 5	△△△△△ △△△△△ 10				△ 1		9
Schoolchild	48	△△△△ △△△△ △△△△ △△△△ △ 21	△△△△△ △△△△△ △△△△△ △△ 17	△△△ △△△ 6	△ 1	△△△ 3			38
Child	8	▽▽▽▽ 4	▽▽▽ 3	▽ 1					10
8	108	48	42	7	5	3	3	102

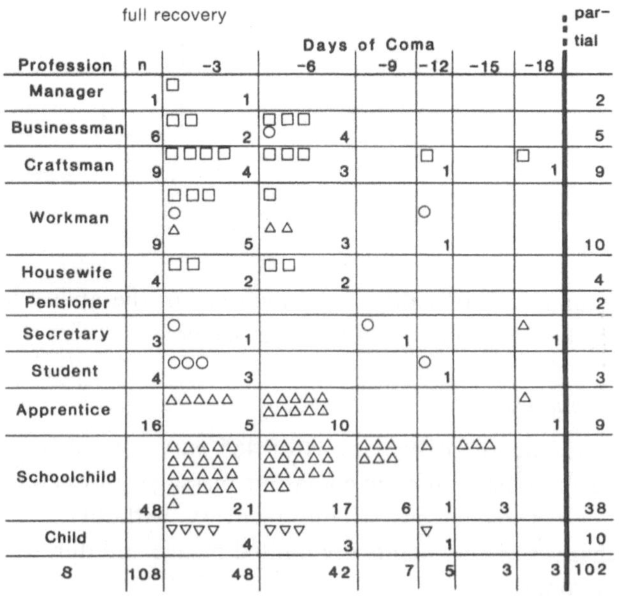

Fig. 4. Vocational resettlement after long-lasting coma: full recovery and partial recovery

Profession

In Fig. 4, 108 patients with normal return to work, that is, one third of survivors of long-lasting coma, are listed according to their pretraumatic activities or profession and duration of coma. More than two thirds of these patients were children, school-children and apprentices up to 20 years of age. Even school-children comatose for 7 to 15 days recovered completely after sufficient rehabilitation.

The last third of the patients is scattered over small groups of students, housewives, workmen and craftsmen. There were only six businessmen and one manager.

The distribution of patients in each professional field was identical in the pretraumatic group, the group of patients with full recovery and the group of patients with partial recovery.

References

1. Brihaye J, Gurdjian ES, Christensen JC, Frowein RA *et al* (1979) Glossary of neurotraumatology. Acta Neurochir (Wien) [Suppl] 25
2. Brooks N (ed) (1984) Closed head injury. Oxford University Press, Oxford, New York, Toronto
3. Edna TH, Cappelen J (1987) Return to work and social adjustment after traumatic head injury. Acta Neurochir (Wien) 85: 40–43
4. Frowein RA, Haar K a d (1987) Rehabilitation after severe head injuries. Advances in Neurosurgery, vol 14: 272–277
5. Jennett WB, Bond MR (1975) Assessment of outcome after severe brain damage. Lancet I: 480–484
6. Poppelreuter W (1937) Psychologische Untersuchungen bei Hirnverletzten. Arch Psychol 98: 279
7. Wieck HH (1977) Lehrbuch der Psychiatrie. 2. Aufl. Schattauer, Stuttgart

Correspondence: Prof. Dr. R. A. Frowein, Neurochirurgische Universitätsklinik, Joseph-Stelzmann-Strasse 9, D-5000 Köln 41, Federal Republic of Germany.

Acta Neurochirurgica, Suppl. 44, 74–77 (1988)

A Relativistic Cybernetic Model for the Personality Disorders Caused by Head Injuries*

I. A. Oprescu and **G. Burstein**

Department of Neurosurgery, Emergency Hospital Bucharest, Rumania

Summary

A mathematical model is used to demonstrate the personality disturbances caused by head injuries. A formal triontic model of personality was developed to express the effects of head injuries. The mathematical system theory (internal model principle) and the relativistic cybernetics are essentially used to explain how severe head injuries destroy the structural homeostasis of personality by influencing the non-cognitive elements via the cognitive elements. Attention is particularly paid to the communication processes between the personality and its external universe.

Keywords: Head injuries; personality disorders; relativistic cybernetics.

1. Introduction

The purpose of this work is to describe mathematically (at a formal, abstract level) the hyperentropising effect of severe head injuries on human personality structure and functioning. Our attempt concentrates on the communication processes in which the personality is involved.

2. A Triontic Formal Model of Personality's Structure

Pamfil proposed a structured model for the human personality[5]. This is illustrated in Fig. 1.

It consists of three interacting structural units making up a triad:

ME represents the self, the structural unit containing the basal features of personality and the psychic energetism.

YOU is the structural unit through which the personality opens toward communication. YOU prepares and strategically supervises the communication of the personality with the external universe.

HIM is the structural unit representing the controller of the interaction ME ↔ YOU. HIM contains an internal model of the external universe.

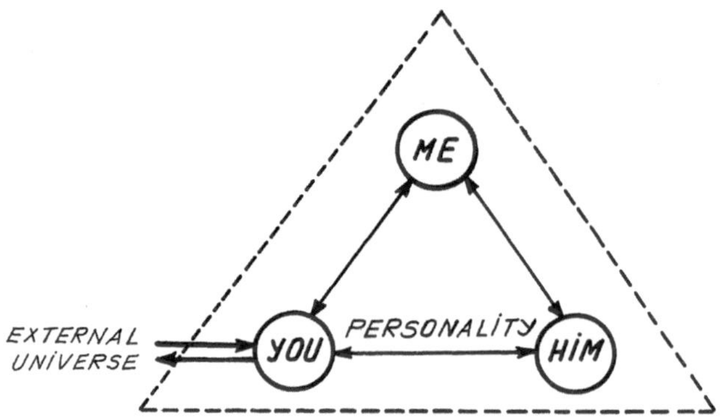

Fig. 1. Triontic formal model of personality's structure (Pamfil)

3. A Theoretic Approach System to Personality

This triontic formal structural model of personality may be reduced, by using the general mathematical system theory, to three feed-back systems. We shall concentrate on the important feed-back system in which HIM controls the interaction ME ↔ YOU subjected to the continuous influence of the external universe. Using the internal model principle from the general system theory[6] we can provide a theoretic explanation system for the importance of the internal model reflecting the external universe contained in the controller unit HIM (Fig. 2).

We will now develop our mathematical theory describing how HIM ensures the homeostasis of the personality, controlling the communication processes between the personality and its external universe. Meanwhile, using relativistic cybernetics[4], we shall develop

* This paper was prepared after discussions with Dr. I. Voinescu, Institute of Neurology and Psychiatry, Bucharest.

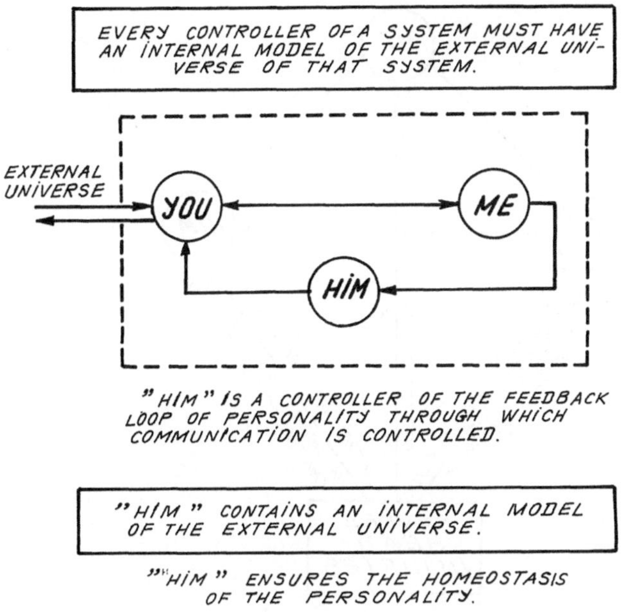

Fig. 2. The internal model principle of the general system theory applied to the personality's feed-back loop having HIM as controller

a mathematical model for Pamfil's formal triontic model of personality described above.

According to relativistic cybernetics, each system is described by four parameters in a four-dimensional space:

H_i — the internal entropy
H_e — the external entropy
V — the transformation potential function
W — the target function.

We remind the reader that the entropy is a measure of the information showing the organization degree (the order degree) of a system. It is related to the balance between order and disorder, organization and chaos etc. While H_i and H_e are rather quantitative parameters (coordinates), V and W represent qualitative parameters. In order to describe the transformations of these parameters by observation and induction we postulate, as relativistic cybernetics does, that the following quantitative-qualitative energy (E) is preserved, as an invariant, in any observation or induction process:

$$E = H_e^2 — H_i^2 — V^2 — W^2.$$

In the four dimensional abstract space considered before, E induces an Einstein-Minkowski metric (dm^2):

$$dm^2 = dH_e^2 — dH_e^2 — dV^2 — dW^2.$$

So that for any system, S, and observers, O and O′, we must have

$$dm^2(S) = dm^2(S \mid 0) = dm^2(S \mid O \mid O')$$

where S / O denotes "S observed (reflected, decoded, etc.) by O" etc. In any observation process, we introduce in addition an observation capacity parameter, $u \varepsilon [0,1]$. Let u be this subjectivity parameter of O in the above abstract example. The equations for obtaining the observed parameters result form the Einstein-Lorentz transformation group which leaves invariant dm^2

$$H_i(S \mid O) = \frac{H_i(S) + u H_e(S)}{\sqrt{1 — u^2}}$$
$$H_e(S \mid O) = \frac{H_e(S) + u H_i(S)}{\sqrt{1 — u^2}}$$
$$V(S \mid O) = V(S)$$
$$W(S \mid O) = W(S)$$

u can be obtained as

$$u = \frac{dH_i(S \mid O)}{dH_e(S \mid O)}$$

and is in fact the result of a dynamic process and not just a parameter.

After these preliminaries from the relativistic cybernetics of Jumarie let us come back to our personality model and let us consider a dynamic abstract system, S, from the external universe, described by its state transition function (dynamics)

$$S \overset{d_s}{\rightarrow} S$$

S can be a physical object, a human being, a concept, a theory etc. and is described as before by

$$S(H_i(S), H_e(S), V(S), W(S))$$

and by a self-observation (self-reflecting) parameter $U_s(S \mid S) = U_s(d_s)$ depending on the dynamics d_s.

HIM is a structurally stable controller of the

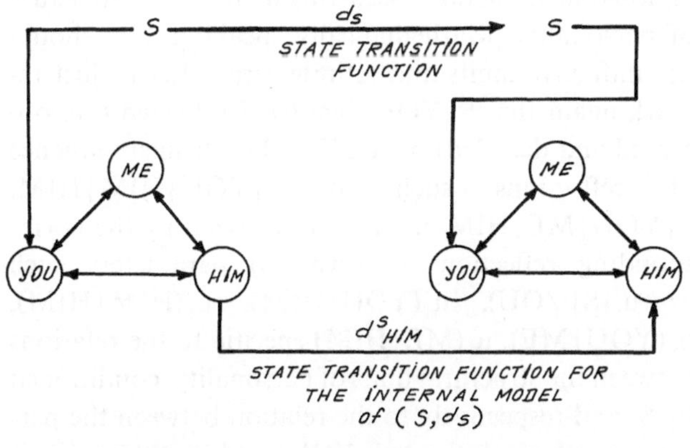

Fig. 3. The algebraic commutative diagram of the internal model principle applied to the triontic model of personality and to the external system (S, d_s)

ME $\leftrightarrow$ YOU interaction only if it incorporates an internal (dynamic) model of the external universe of the personality so also of (S,d; $_s$). According to the abstract internal model principle the following diagram must be commutative, in the algebraic sense (Fig. 3).

This diagram shows how internal models are obtained in HIM. We obtained the following mathematical conditions (Σ):

$$\mathscr{H}_i^S(\text{HIM}) = H_i(S|\text{YOU}|\text{HIM}|\text{HIM}) + H_i(S|\text{YOU}|\text{ME}|\text{HIM}) = H_i(S|S|\text{YOU}|\text{HIM}) + H_i(S|S|\text{YOU}|\text{ME}|\text{HIM})$$

$$\mathscr{H}_e^S(\text{HIM}) = H_e(S|\text{YOU}|\text{HIM}|\text{HIM}) + H_e(S|\text{YOU}|\text{ME}|\text{HIM}) = H_i(S|S|\text{YOU}|\text{HIM}) + H_i(S|S|\text{YOU}|\text{ME}|\text{HIM})$$

$$V(S|\text{YOU}|\text{HIM}|\text{HIM}) + V(S|\text{YOU}|\text{ME}|\text{HIM}) = V(S|S|\text{YOU}|\text{HIM}) + V(S|S|\text{YOU}|\text{ME}|\text{HIM})$$

$$W(S|\text{YOU}|\text{HIM}|\text{HIM}) + W(S|\text{YOU}|\text{ME}|\text{HIM}) = W(S|S|\text{YOU}|\text{HIM}) + W(S|S|\text{YOU}|\text{ME}|\text{HIM})$$

$$dm^2(S) = dm^2(S|\text{YOU}|\text{HIM}|\text{HIM}) = dm^2(S|S|\text{YOU}|\text{HIM}) = dm^2(S\,\text{YOU}|\text{ME}|\text{HIM}) = dm^2(S|S|\text{YOU}|\text{ME}|\text{HIM})$$

where $\mathscr{H}_i^S$ (HIM) and $\mathscr{H}_e^S$ (HIM) are respectively the total internal and the total external entropy induced in HIM by S.

Using the aforementioned theoretical background these synthetic formulae can be fully developed replacing the abstract observer O by ME, YOU and HIM. We shall not insist on this. One may remark that these formulae are just a first order additive approximation of the infinite iteration of reflections, observations and inductions occurring in the triadic structure of personality. Mathematically speaking, one should use inductive limits and infinite series. Let us just remark again that S | YOU denotes S reflected (*i.e.* observed and decoded) by YOU and each such sequence of reflections such as S | YOU | HIM | HIM, S | YOU | ME | HIM etc. is characterized by the corresponding reflection (observation) parameters such as u_s(S | YOU), u_s(YOU | HIM), u_s(HIM | HIM), u_s(YOU | ME), u_s(ME | HIM) specific to the relations between the structural units of personality, conditioned by S, and respectively to the relation between the personality (through the unit YOU) and its external universe (symbolized by S). As one can see from the preceding diagram (Fig. 3) u_s (HIM | HIM) actually generates d^S_{HIM}, the state transition function for the internal dynamic model of (S, d_s). We repeat again that all these reflection or observation or decoding parameters are in fact outputs of dynamic interaction mechanisms.

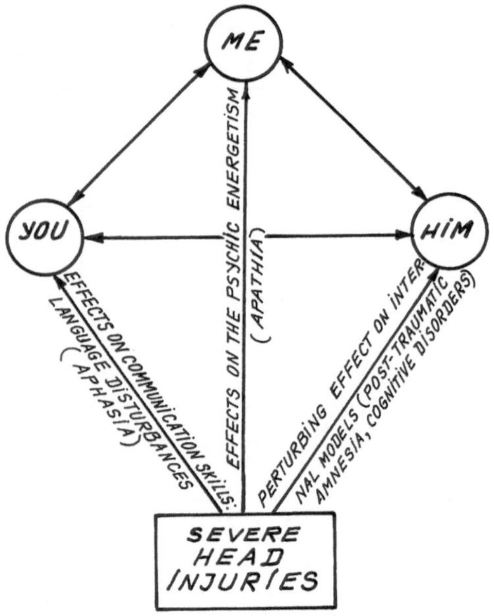

Fig. 4. The effects of severe head injuries on the personality's triadic structure

4. Head Injury Effects on Personality: a Mathematical Approach

We can now approach head injuries disturbance of personality in a mathematical framework. Based on the well-known data concerning head injury effects on personality (§ 12, 3) we can show a diagram (Fig. 4) in which the effects of severe head injuries are decomposed according to the three directions corresponding to the three structural units of personality.

We can now express mathematically the hyperentropising effect of head injuries, *particularly exerted on the structural unit of personality HIM*, by considering again the internal model principle diagram in the presence of head injuries (Fig. 5).

As mentioned earlier, perhaps the strongest effect of severe head injuries is against the HIM unit by post-traumatic amnesia and cognitive disorders that disturb the internal models contained by the controller HIM, by increasing in it the entropy (disorder, lowering of the organization degree), and finally breaking the homeostasis of the personality, particularly in its communication processes with the external universe (exterior world). Returning to the system of equations (Σ) we

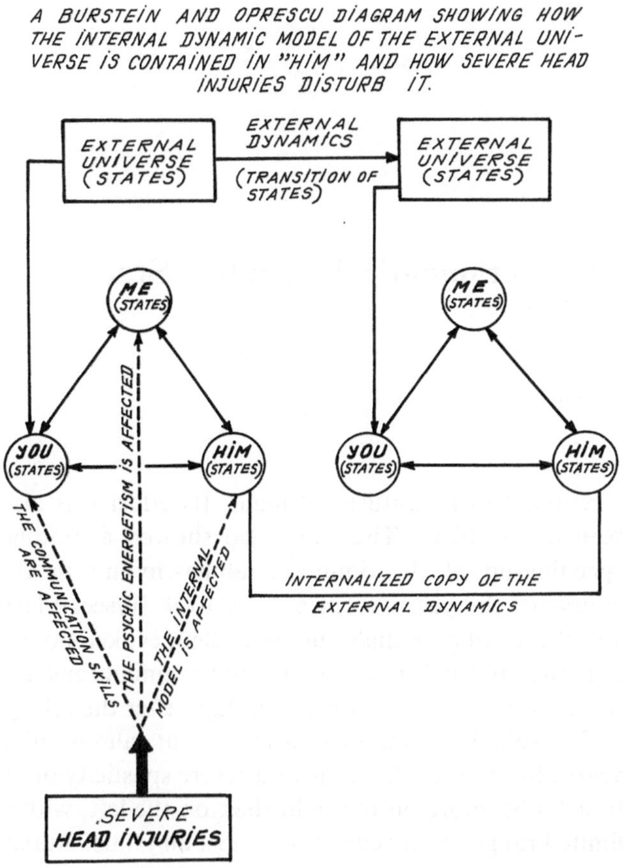

Fig. 5. An algebraic commutative diagram showing how the internal dynamic model of the external universe (contained in HIM) is disturbed by severe head injuries

can see that the increase of entropy in HIM, due to head injuries (disorders):

$$\mathscr{H}_i^S(\text{HIM}) = \mathscr{H}_i^S(\text{HIM}) + \mathscr{H}_i^{\text{additional}}$$
$$\mathscr{H}_e^S(\text{HIM}) = \mathscr{H}_e^S(\text{HIM}) + \mathscr{H}_e^{\text{additional}}$$
$$(\Gamma)$$

breaks the homeostasy represented by the preservation of the invariant $dm^2(S)$ (quantitative-qualitative energy metric) as now due to the equations (Γ):

$$dm^2(S) < dm^2\ \overline{(S \mid YOU \mid HIM \mid HIM)}$$
$$dm^2(S) < dm^2\ \overline{(S \mid S \mid YOU \mid HIM)}$$
$$dm^2(S) < dm^2\ \overline{(S \mid YOU \mid ME \mid HIM)}$$
$$dm^2(S) < dm^2\ \overline{(S \mid S \mid YOU \mid ME \mid HIM)}$$

where the bar over $\mathscr{H}_i^S(\text{HIM})$ $\mathscr{H}_e^S(\text{HIM})$, S | YOU | HIM | HIM etc. denotes the effect of a severe head injury and "<" (means "smaller than" which also implies "not equal to").

5. Conclusions

We particularly insisted on HIM because of its role in the personality structure and also because of the more frequent and more serious disturbances caused by head injuries to HIM, but as can be seen in our formulae, the multiple interactions between ME, YOU and HIM were not neglected at all. HIM was just the point of view we have chosen to look at personality's triadic structure in which all three units are equally important and make up an entity. As we were more concerned with personality's homeostasis in the communication processes with its external universe, we stressed more the way in which head injuries affect the non-cognitive elements of personality by means of the cognitive ones (*i.e.* ME by means of HIM).

We essentially used here the relativistic cybernetics which is now the most general theory that can be developed for observation processes in human systems. A generalization of the relativistic cybernetics was obtained in[1] using sophisticated elements of differential geometry and differential algebra. Finally, an extension of this work to the effects of head injuries on the personality from a neurophysiological point of view might benefit from the theory developed in[2].

Acknowledgement

The present version of this paper was prepared by G. Burstein, Dr. I. Voinescu and by the care of Dr. Elena Oprescu as a tribute to the memory of late Dr. I. A. Oprescu.

References

1. Burstein G, Voiculescu N (1987) A general theory for systems transformation by observation. Application to human systems; 11th Symposium on Organization and Management, Bucharest
2. Burstein G, Nicu MD, Bălăceanu C (1987) Simplical differential geometric theory for language cortical dynamics, Fuzzy Sets and Systems Journal (special issue on modelling language and cerebral processing) 23: 303–313
3. Jennett B, Teasdale GM (1982) Management of head injuries. FA Davis Co, Philadelphia
4. Jumarie G (1985) Principles of relativistic cybernetics. Gordon and Breach, London
5. Pamfil E, Ogodescu D (1976) Personality and becoming. Editura Ştiinţifică si Enciclopedică, Bucharest
6. Wonham WM (1976) Towards an abstract internal model principle. IEEE Trans Sys Man and Caber, SMC 6, 11: 735–740

Correspondence: Prof. Dr. F. Loew, Neurochirurgische Universitätsklinik, D-6650 Homburg/Saar, Federal Republic of Germany.

Acta Neurochirurgica, Suppl. 44, 78–79 (1988)

Unexpected Improvement After Prolonged Post-Traumatic Vegetative State

W. F. M. Arts, H. R. van Dongen, and **J. Meulstee**

Erasmus University Rotterdam, Institute of Neurology, Rotterdam, Netherlands

Summary

Presentation of an unusual case of severe head injury in which, after three and a half years of vegetative state, a gradual return to consciousness and personality occurred.

Keywords: Head trauma; vegetative state; recovery.

Recovery of consciousness after a long-lasting vegetative state remains a very great exception, as the investigations of Higashi *et al.* (1977, 1981) have definitely shown. Here, we present the unique case of a girl remaining in a vegetative state for three and a half years after severe cerebral trauma and then showing a gradual return of consciousness and personality. On the Glasgow Outcome Scale, she moved up from vegetative state to severe disability. Her degree of residual handicap now is determined by her intellectual and emotional regression (see below), and by severe spasticity and multiple joint contractures with para-articular calcifications.

Immediately after the accident, the GCS was ElM3 Vl. She had abnormal flexion posturing of both arms and tonic extension of the legs. Immediate CT-scanning only showed signs of increased intracranial pressure but after 24 hours, there were bilateral frontal contusions. After three weeks, she gradually moved into a vegetative state. There was no contact nor any sign of a self-conscious existence. Reflex turning of the head toward light or noise was observed, as was reflex grimacing with tachypnoea and tachycardia upon painful stimuli. After three and a half years, she began to react to her surroundings, and eighteen months later, she recognized familiar persons and her personal belongings and was able to take an active interest in them.

The CT-scan then showed gross brain stem and cerebellar atrophy; the cerebral atrophy was largely limited to both temporal lobes. A linear cleft in the white-matter of the left frontoparietal lobe was seen as a remnant of a contusional focus. Its effect was also seen in the EEG. The EEG also showed a gradual speeding-up of the dominant alpha-rhythm in the course of the years, from 6–7/sec to 9–10/sec. Flash visual evoked potentials and brain stem evoked potentials were normal. Evoked potential examinations had not been performed in the acute stage after the injury.

Neurological examination at present, almost nine years after the accident, shows a severe spasticity of all four limbs, more on the right than on the left, with a limited range of movement in almost all joints. Ataxia is not a conspicuous feature. There is a severe, but improving, impairment of mouth and tongue movements, in which apraxic and pseudobulbar mechanisms probably combine. No visual field defects can be found.

Neuropsychological examination: on a standardized Dutch intelligence test, she obtained an IQ of 74. Immediate recall is severely disturbed. On a serial word learning test she failed completely. Memory for remote events is less severely disturbed. For example, the girl could give relevant data about her earlier school experiences, but she did not know her former profession or the name of the office with which she was connected before the accident.

In contrast to the memory functions the improvement of speech appears to be continuing. Nine years after the accident, long utterances can be heard. The articulation of phonemes is not severely defective any more with the exception of the letters K, L, S and T. However, weakening of vowels and consonants occurs after a 5–10 minute conversation in association with increasing dysphonia and irregular audible inspiration. She can repeat sentences of seven words and these sentences are intelligible.

Despite the severe motor handicaps, described above, two important actions (for her) are possible: smoking cigarettes and drinking coffee. The speed of action is slow and she manipulates the object (cigarette-lighter and cup) with the utmost difficulty but she enjoys the cigarettes as well as the coffee.

As regards her personality, her parents described her before the accident as an intelligent girl who attended a secondary school without problems and afterwards was trained as a typist. She was a gay, sometimes excited and quick-tempered girl. However, severe behaviour problems never occurred. At present, parents and nursing staff characterize her as a person who has reasonable social contacts with the other patients. Frequently she attracts attention to herself. If the nursing staff does not pay attention to her, she sometimes starts to cry and shout. She is interested in music and TV-shows and reads books, mostly simple love stories. However she cannot retain the content. Striking is the care for the make-up of her face and hair. She regularly visits a hairdresser.

We do not know whether she has reached a final stage yet. In addition we have two considerations:

1. Although there is little or no improvement in her cognitive functioning, there is considerable progress in her social behaviour and speech. Perhaps the quality of life will ameliorate still.

2. There is a mild shrinking of the retrograde amnesia. We do not know what will happen when she becomes aware of important events, for example the death of her loved one. Perhaps this will change her mood and – in her view – diminish the meaning of her life.

References

1. Arts WFM, van Dongen HR, van Hof-van Duin J, Lammens E (1985) Unexpected improvement after prolonged post-traumatic vegetative state. J Neurol Neurosurg Psychiatry 48: 1300–1303
2. Higashi K, Sakata Y, Hatano M *et al* (1977) Epidemiological studies on patients with a persistent vegetative state. J Neurol Neurosurg Psychiatry 40: 876–885
3. Higashi K, Hatano M, Abiko S *et al* (1981) Five-year follow-up study of patients with persistent vegetative state. J Neurol Neurosurg Psychiatry 44: 552–554

Correspondence: H. R. van Dongen, M.D., Erasmus University Rotterdam, Institute of Neurology, Room EE 2287, Postbus 1738, NL-3000 DR Rotterdam, Netherlands.

Acta Neurochirurgica, Suppl. 44, 80–83 (1988)

Personality Traits After Prolonged Vegetative State

Reflections on "Unexpected Improvement After Prolonged Post-traumatic Vegetative State" by W. Arts, H. van Dongen, J. Meulstee

E. Thiery

Department of Physiological Psychology and Neuropsychology, State University of Ghent, Belgium

Summary

The consequences of the trauma on the personality traits of the case described by Arts *et al.* are analysed. Among the biological factors which favoured the restoration of consciousness, youth, normal pretraumatic personality as well as the location and extent of the brain lesions have been of special importance.

The psychopathological changes mainly consisted in an impairment of higher mental functions with intellectual regression, loss of abstraction ability and a severe deficit of recall and memory.

The pragmatical and ethical consequences of this case are discussed.

An arresting and unique case of unexpected improvement after prolonged post-traumatic vegetative state is published in this book (Arts *et al.* 1985) and was presented with illustration by a videofilm at the Academia.

We want to reflect on the consequences of the trauma on the personality traits of this woman who very unexpectedly regained consciousness, self-experience and communication.

In the first place we will discuss the biological, psychological and social factors that may have contributed to the fact that she regained consciousness.

In the second place we will comment on the psychopathological mechanisms that determine the modified post-traumatic personality.

In the third place we will take up some of the socioeconomic and ethical considerations raised by this most interesting communication.

Firstly some comments on the biopsychosocial factors that contributed to a most remarkable and uncommon clinical evolution with restoration of a conscious existence after two and half years of a vegetative state, that literature (Jennett and Plum 1972) has come to call a persistent vegetative state. Persistent it almost always appears to be after six months or one year of vegetative state (Minderhoud and van Zomeren 1984), notwithstanding some extremely rare cases of late but limited mental recuperation as mentioned by Higashi *et al.* (1977, 1981), Rosenberg *et al.* (1977) and Bricolo *et al.* (1980).

The biopsychosocial factors that may influence the clinical evolution of personality restoration can be categorized in pre-traumatic, traumatic and post-traumatic factors although these factors are always strongly mingled and intensely interacting (Walsh 1985). Looking at them in a systematic way nevertheless makes sense (Thiery *et al.* 1982). In the present case looking at them may be helpful in coming to understand a little bit better what may have contributed to the unexpectedly positive evolution of the case.

As a pre-traumatic factor *age* certainly played a positive role. The chances of post-traumatic recuperation become less with progressing age (Bond 1975). This is caused by a reduction of functional reorganization of equipotential structures and systems and is caused by a diminished regeneration on the axonal level, as shown by Jeannerod and Hecaen (1979). In the study of Bricolo *et al.* (1980) about prolonged post-traumatic unconsciousness one can clearly see that the average age of the patients with a better prognosis was nineteen years, and that the chances of any recuperation clearly fell with progressing age. Age appears to be a very important factor not only for the survival ratio, but also for the quality of recuperation. At eighteen years of age the situation was most favourable for this patient.

As pretraumatic, favourable factors for a more propitious evolution we can also mention her normal *physical* and normal *intellectual* status, as well as her normal *premorbid personality*, as premorbid personality problems play a role in the appearance in post-traumatic personality of psychotic, depressive and psychopathic traits (Klonoff 1971).

As a traumatic factor the exact nature of the *aetiology* must be taken into account. Here the pathological state is related to a closed head injury. This was a severe injury, causing immediate deep coma (with $E_1M_3V_1$ on the Glasgow Coma Scale), but nevertheless an injury that did not require any neurosurgical intervention (such as for a haemorrhage), and an injury that did not provoke direct complications (such as acute hemisphere swelling). It also has to be mentioned that spontaneous breathing was restored quite quickly, as she was able to breathe independently after twenty four hours. The evolution from deep coma to a state of "wakefulness without awareness" already came in the first three weeks, but then remained unchanged for the following 30 months.

In our opinion very important traumatic factors for the unexpectedly positive, late recuperation in the level of awareness, personality and communication have been the *location* and *extent* of the lesions.

On the first CT scan there was subarachnoid blood, on a later scan there appeared to be some frontal contusion and on the latest CT scan there is infratentorial atrophy seen in combination with, on the supratentorial level, only some temporal atrophy and a left fronto-parietal linear white matter hypodensity.

With this pattern this patient falls (in Lobato's anatomical categorizing system) into a group with a clearly better prognosis (Lobato *et al.* 1983). As Plum and Posner (1980) stated, the bad prognosis of a vegetative state is caused by its pathologic basis. They said "the pathologic basis of the vegetative state usually consists of badly damaged cerebral hemispheres combined with a relatively intact brain stem". As Minderhoud and van Zomeren (1984) state the hemisphere lesions are usually extensive white matter lesions. Whereas when there is extensive necrosis at the level of the grey matter around the aqueduct and the third ventricle a state of prolonged coma results from a "protracted post-traumatic encephalopathy" as described by Jellinger and Seitelberger (1970).

In this case the situation was quite the opposite of what Plum and Posner (1980) describe, with an important sparing of the substance of the hemispheres not only of the white but also of the gray matter. This makes the evolution more understandable. Once the reticular ascending, activating system in its brain stem portion had regained enough functional recuperation and reorganization, higher especially cortical structures could once again be reactivated thanks to the dynamic reticulo-thalamo-cortico-frontal interplay made possible by the reticulo-cortico-reticular loops restoring some cognitive, personality and communication activity in this brain after many months of total silence. The unique location and extent of lesions in this case presumably made possible a late and partial neocortical reactivation after a long brain stem restoration.

The electro-encephalographic, evoked potential and neuro-ophthalmic results are also an argument in favour of the restored and/or reactivated functional capacities of a large cerebral area.

As a post-traumatic factor for positive evolution the absence of serious *late complications* certainly played a role (no normotensive hydrocephalus, post-traumatic epilepsy, and so on). It is also important that the medical and paramedical treatment, help, attention and stimulation have been early, long-lasting, intense and adequate. This certainly improved emergent mental properties.

On the other hand physical therapy could not stop tetraspasticity and contractures, the most important reason why the patient remained and will remain with a "severe disability", on score 3 of the Glasgow Outcome Scale.

All in all a remarkable evolution but on the whole a sad story.

As for the biopsychosocial factors that favourably influenced the evolution we stated that a positive constellation was present with, as the most important elements, the patients youth and the site and extent of the causative lesions.

The psychopathological changes in personality after trauma, with prolonged coma or vegetative state are impressive.

Firstly there is the impairment of higher mental functions in a context of intellectual regression (Thiery 1986). This previously normally intelligent girl now functions at an IQ level in the mid 70s. Gnosia, phasia, and lexia are grossly well preserved while immediate recall and memory, especially short-time memory are very defective. This constellation of preserved versus defective higher mental functions is neuropsychologically typical for a post-traumatic state and easily understandable in the light of the lesions preferentially located in deep cerebral, centrencenphalic, reticulo-limbic structures.

Besides these deficiencies on the intellectual level and on the higher mental or psycho-instrumental functions we always note a loss of abstract attitude after head injury (Goldstein 1952). A loss of abstract attitude caused by the disintegration of the brain as a unified, globally functional total structure.

Furthermore some traits of the post-traumatic personality are the reflection of ego defensive autoprotective mechanisms at work. Compulsive attitudes, loss of interest, apathia, agression, and catastrophic reaction and partially also anosognosia, amnesia, confabulation and perseveration have to be understood in that way.

Didn't nature do a good thing by hiding from the young girl the fact that she had lost her fiancé in the accident? We can follow Goldstein who interprets some of the personality traits of the brain-injured patient as mechanisms of defence against a dramatically lost psychomental equilibrium.

Besides all this there is also the ever present threat of neurotic or psychotic regression, and especially the threat of depression as a result of psychodynamic, organic and social pathological interaction.

This girl is but a shadow of her former self. Her loss of intellect, her attention and memory deficit, her enormous physical handicap made her terribly handicapped. Thanks to her autoprotective ego defensive mechanisms a total personality disintegration is prevented. She lives courageously, she sticks gamely to her past, fighting hard for a sad life, remaining forever dependent on family, the medical staff and the good will of society.

What can we learn from this case on the pragmatic and ethical level? Doctors, lawyers and philosophers come to the growing insight that we have to work urgently on renewed, adapated and refined visions and points of view on the exact role medicine has to play for humanity.

In former ages the doctor was rather impotent, rather powerless; his one and only aim was to preserve life, to fight for the life of the patient.

In the last decades the doctor has become very powerful, thanks to an ever growing technology. He has now to reconsider and to refine his ethical view-points, in the knowledge he has not only to serve life but that he has to bring the patient a quality of life. He has to fight for a life worth living. If he is not able to offer the patient a respectable life then he has to renounce the battle, otherwise driven by "hubris" or presumptuousness he lacks respect for the patient and the patients right to a respectable death.

As Jennett (1976) puts it many statistics show that there is a growing awareness among the general public that a combination of mental with physical handicap makes disability after brain damage devastating for the patient and the family, and causes a disaster bigger than death itself.

The present case of unexpected improvement after prolonged post-traumatic vegetative state shows us that in medecine for every rule there is an exception. This must prevent us from a thoughtless, inconsiderate euthanasia. This also forces us to still more statistical studies on outcome after severe brain damage.

The present case teaches us also that secondary phenomena (such as the problem of contractures) can dramatically hamper the positive mental evolution.

But above all the present case teaches us that even the exception to the rule still remains a human disaster of severe disablement and painful, total dependency.

Nature does not know joy, happiness and felicity. Man has to build them. So he must be able to build them, and it is the humble role of medicine to try to make him able to do so.

We must not forget this girl while setting goals in our everyday practice and in our ethical considerations.

References

1. Arts W, van Dongen H, van Hof-van Duin J, Lammens E (1985) Unexpected improvement after prolonged post-traumatic vegetative state. J Neurol Neurosurg Psychiatry 48: 1300–1303
2. Bond M (1975) Assessment of the psychosocial outcome after severe head injury. In: Porter R, Fitzsimmons D (eds) Outcome of severe damage to the central nervous system. Elsevier, Amsterdam, pp 141–157
3. Bricolo A, Turazzi S, Feriotti G (1980) Prolonged post-traumatic unconsciousness. J Neurosurg 52: 625–634
4. Goldstein K (1952) The effects of brain damage on the personality. Psychiatry 15: 245–260
5. Higashi K, Hatano M, Abiko S *et al* (1981) Five years follow-up study of patients with persistent vegetative state. J Neurol Neurosurg Psychiatry 44: 552–554
6. Higashi K, Sakata Y, Hatano M *et al* (1977) Epidemiological studies on patients with a persistent vegetative state. J Neurol Neurosurg Psychiatry 40: 876–885
7. Jeannerod M, Hecaen H (1979) Adaptation et restauration des fonctions nerveuses. Simep, Villeurbanne, pp 273–314
8. Jellinger K, Seitelberger F (1970) Protracted post-traumatic encephalopathy. J Neurol Sci 10: 51–94
9. Jennett WB (1976) Resource allocation for the severely brain damaged. Arch Neurol 33: 595–597
10. Jennett WB, Plum F (1972) Persistent vegetative state after brain damage. Lancet 1: 734–737
11. Klonoff H (1971) Head injuries in children. Am J Public Health 61: 2405–2417
12. Lobato R, Cordobes F, Rivas J *et al* (1983) Outcome form severe head injury related to the type of intracranial lesion. J Neurosurg 59: 762–774
13. Minderhoud J, van Zomeren A (1984) Traumatische hersenletsels. Bohn, Scheltema & Holkema, Utrecht, pp 104–107

14. Plum F, Posner J (1980) The diagnosis of stupor and coma. Davis, Philadelphia, pp 5–6
15. Rosenberg G, Johnson S, Brenner R (1977) Recovery of cognition after prolonged vegetative state. Ann Neurol 2: 167–168
16. Thiery E (1986) Stoornissen van het persoonlijkheidsprofiel in het licht van hersentraumata. In: Bakelmans R, Verhofstadt-Deneve L (eds) Ontwikkeling, persoonlijkheid en milieu. Acco, Leuven, pp 401–411
17. Thiery E, Dietens E, vander Eecken H (1982) La récupération spontanée: ampleur et limites. In: Seron X, Laterre S (eds) Rééduquer le cerveau. Margada, Bruxelles, pp 33–43
18. Walsh K (1985) Understanding brain damage. Churchill Livingstone, Edinburgh, pp 144–183

Correspondence: Prof. Dr. E. Thiery, H. Dunantlaan 2, B-9000 Gent, Belgium.

Acta Neurochirurgica, Suppl. 44, 84–92 (1988)

Two Cases of Painters Operated More than Ten Years Ago for Intracranial Lesions. Evolution of Their Artistic Production

G. Foroglou, G. Kaprinis, and **C. Phocas**

Neurosurgical Clinic of the University, AHEPA General Hospital, Thessaloniki, Greece

Summary

Two female professional painters, who underwent a neurosurgical operation, are described. Their artistic production before and after the neurosurgical intervention is compared and the observed modifications are discussed.

Keywords: Neuropsychology; intracranial lesion; artistic production.

Introduction

Neuropsychological literature contains many studies regarding the influence of brain lesions on colour-perception and its disturbances (Kleist 1934, Faust 1955, Hécaen and Angelergues 1963, 1965, Geschwind 1965, de Renzi and Spinnler 1967, Lhermitte *et al.* 1969, Oxbury *et al.* 1969) as well as on the manipulation of spatial data (Brain 1941, Critchley 1953, Battersby *et al.* 1956, Hécaen *et al.* 1956, Costa *et al.* 1969, Leicester *et al.* 1969, de Renzi *et al.* 1970, Hécaen 1972 a, b). At the same time, one can find many examples of writing and drawings produced by patients who were non-professional painters. This last point was what particularly urged us to present the two cases of professional painters who underwent neurosurgical operation and to investigate the evolution of their clinical condition in relation to their artistic production.

History

The first patient. Aged 29 years, right-handed, was admitted to our clinic for intracranial hypertension, with motor and sensory impairment on the right side. The pre-operative investigations showed a huge space-occupying lesion, which at operation was found to be an epidermoid extending from the frontal pole back to the occipital lobe (Fig. 1). Neurosurgical progress has been uneventful.

During the first post-operative neuropsychological examinations, she showed dysarthria, right hemiparesis and the dichotic listening test resulted in a clear superiority of the right ear although the patient was right-handed.

Colour perception was investigated by different methods.

The Farnsworth test of 100 Hue*, resulted in the conclusion that there was a disturbance in colour perception of tritanopic type – namely on the blue-yellow axis. During retesting, which took place three months later, the only pathological findings still existing, were the tritanopic type disturbances in colour recognition, although these findings disappeared on retesting at five and eight months post-operatively.

From a psychiatric point of view, during the first sessions, there was to a certain extent an observable depression in relation to her health condition, obviously reactional in type although these symptoms were greatly diminished eight months after the operation.

At present, her neuropsychological test results seem to be in normal limits, as well as her MMPI profile. The psychiatric evaluation shows a normal person, with no major psychological problem and she is able to organize her life and interpersonal relations in a self-sufficient satisfactory manner.

The second patient, aged 50 years, also right-handed, was operated on for a huge acoustic neurinoma on the right side. The CT-scan showed compression of the cerebellum and brain stem without any sign of hydrocephalus. A post-operative right-sided facial nerve paresis cleared up a few months later.

Today, she is found to be free of neurological symptoms but from the psychiatric point of view she shows exaggerated behaviour in many dimensions. However, as she was generally quite co-operative she asked us, in an obviously anxious way, for confirmation that she was not contradicting our expectations. Moreover, being hyperkinetic and active, she showed a very hypomanic picture, although the pleasant emotional state of the condition was missing.

* The Farnsworth-Munsell 100 Hue Test for the Examination of Colour Discrimination (Munsell Color Co, Baltimore) is composed by 85 movable buttons, representing colour examples from the Munsell Atlas of colours placed in four different boxes. The movable buttons are placed randomly and the four boxes are presented to the patient successively. The patient is asked to place the movable buttons in such a way that they represent the best colour gradation between two fixed colour buttons. Concerning the acquired dyschromatopsies the most frequently observed traces are those between tritanopic and deuteranopic type. Finally, this test is a very sensitive tool for the colour discrimination and permits qualitative notions for acquired colour dyschromatopsies (Franceschetti *et al.* 1963).

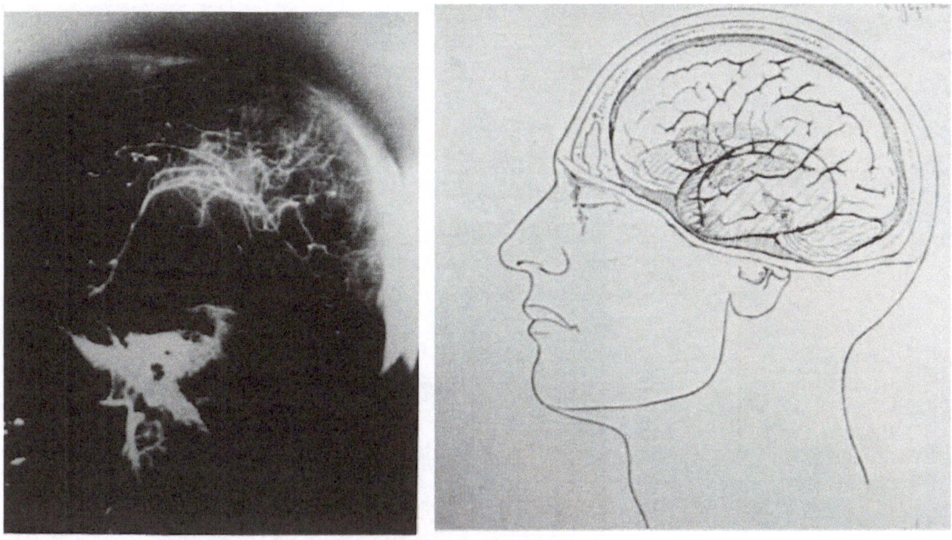

Fig. 1

Artistic Production Analysis

First Case

The painter's artistic work started with black and white pieces (Fig. 2) in a tonal-gradient painting. She later proceeded using colour with significant addition in the first place of black or grey (Fig. 3). A little later, she started using blue, violet and yellow colours (Figs. 4, 6), in which she presented a certain difficulty in the Farnsworth test.

Her design, unstable in the first place (Fig. 5), become more organized and more robust (Figs. 3, 6).

We believe that what she herself stated is of great importance. She said that her work was no more limited to its subject but she could envision it expanding, as a composition, beyond the limits of the canvas or the definite space of the painting.

Second Case

Our second case, with nine personal exhibitions and participation in 15 group exhibitions seems to be a

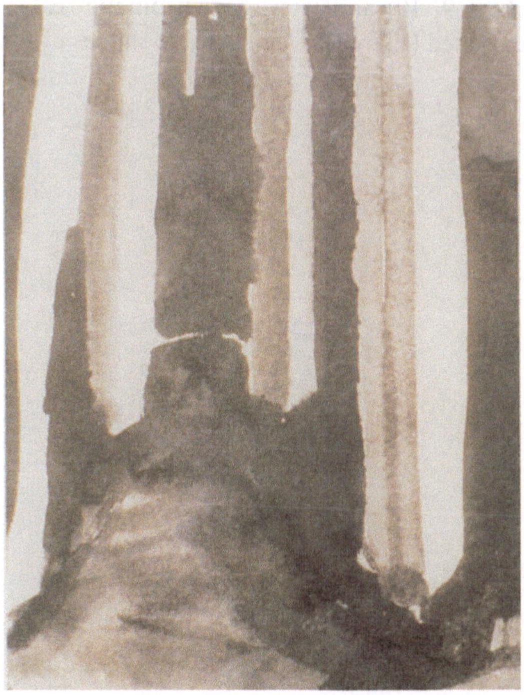

Fig. 2

Fig. 3

Fig. 4

Fig. 6

peculiar type of painter. The materials which she uses in her work are taken from nature, such as leaves and flowers collected by herself at different seasons of the year. In the beginning, the artist tried to match her materials in a two-dimensional arrangement (Figs. 7, 8).

Immediately after the operation, a considerable difference is observed. Her paintings escape two dimensions (Fig. 8) and conquer the third one (Fig. 9).

Concerning her materials, the leaves and flowers (Figs. 7, 8) are dissolved in order to enable the artist for more comfortable manipulations (Fig. 10), although they maintain their nature. Later, they are cut into pieces, so that they can no longer be recognized in their primary condition (Figs. 11–17).

As regards the colour used during the first period, where the most decorative paintings had the colour of her flowers (Figs. 7, 8), she passes to shades of brown (Figs. 9, 10) and later, her chromatic scale is greatly enriched (Fig. 11). Usually there shades of blue appear

Fig. 5

Fig. 7

(Fig. 12) while black appears in a threatening way (Fig. 13).

At the same time, her expressional ability is enriched by a new element, namely the darkness (Fig. 14) and the light (Fig. 15) that the artist starts to control effectively.

Concerning the choice of her subjects, she started from bouquets (Fig. 7) before the operation. Postoperatively she progressed to brown trunks (Fig. 9) and landscapes (Fig. 10), more or less concrete, to continue with mixing of blue and black tones, in rocks viewed closely (Fig. 16) and finally, with petrified faces (Fig. 13), figures of more or less tragic appearance (Fig. 17).

Discussion

Although neurosurgical operations and modifications of artistic production cannot be correlated in an absolute cause-and-effect manner, the presentation of these two cases permits certain assumptions to be made.

Our first comments concern the tritanopic type disturbances in colour perception of our first case. We agree with Lhermitte (1972) that Farnsworth 100 Hue test is the proper instrument for the detection of these disturbances. Actually no other examination gave even a minimal indication of a difficulty in colour perception

Fig. 8

Fig. 9

Fig. 10

Fig. 11

Fig. 12

Fig. 13

in our first case. Lhermitte characteristically says that tritanopic disturbances are not often reported because the Farnsworth test is not being used in cerebral diseases.

As to the localizing value of the symptom there is a general acceptance that the deficiencies in colour recognition should be attributed to left hemisphere disturbances, while only de Renzi *et al.* (1969) report a higher incidence of colour perception disturbances in right hemisphere lesions using the Ishihara test. Lhermitte *et al.* (1967) claim that only patients with posterior bilateral lesions present colour perception disturbances, which can be easily detected by all relative tests (Ishihara, distinction, matching). For all other patients they suggest the Farnsworth test, especially for the tritanopic type disturbances. These disturbances represent an elementary dysfunction of colour vision caused by cortical-subcortical lesions in the specific information pathways of the blue-yellow axis. Such disturbances may remain residual and are most often observed in temporal lesions.

A second noticeable element mostly in the second case is the modification in the facility of controlling the spatial data, where the passage from two dimensional arrangement to three-dimensional synthesis was convincingly demonstrated. According to Hécaen *et al.* (1956) disturbances in the spatial data manipulation represent an apractic-agnostic syndrome of the posterior "carrefour" of the non-dominant hemisphere (in fact, the operation on the second case involved the right cerebellar region). Nevertheless, it should be recalled, that as our first patient herself reported, her work was no more limited to its subject but she could envisage it expanding in a composition beyond the canvas' limits of the definite space of the painting. Certainly, this

Fig. 14

Fig. 15

could be irrelevant to the ability of operating spatial facts. It could be connected to other more general potentials of the CNS, such as the integration of a picture, the extending of a form, etc. In this case we should refer to the integrating action of the tertiary zones of the parieto-temporo-occipital cortex (Luria 1973).

It appears that apart from partial functions like colour perception or functions related to spatial facts, where a localizing value could be attributed, there are more general qualities such as subject conception, ability for composition, enrichment of expression, which represent the function of larger areas or even the function of the brain as a whole.

It should be noted that in both cases the neurosurgical operation did not result in any removal of brain tissue but on the contrary, in the removal of the space-occupying element which disturbed the brain tissue, that regained its normal functional conditions postoperatively.

One might say that in our cases the Lashley mass

Fig. 16

Fig. 17

effect (Lashley 1929) operated in favour of our patients since a significant part of the brain was relieved. In the area of the artistic production of our patients this relief was accompanied by the liberation of new creative capabilities.

Conclusions

1. As regards our first patient we conclude that the application of the Farnsworth 100 Hue test allowed the detection of tritanopic type disturbances in colour perception. These disturbances are considered to be correlated with left hemisphere lesions.

2. As regards out second case we conclude that modification in the facilitiy of controlling the spatial data represents an apractic-agnostic syndrome of the posterior "carrefour" of the right hemisphere.

3. As a final concluding remark we notice that apart from partial functions (like colour perception etc.), which can be related to specific cortical areas, there are more general qualities such as subject conception, ability for composition and enrichment of expression that represent the function of larger cortical areas or even the function of the brain as a whole.

References

1. Brain R (1941) Visual disorientation with special references to the lesions of the right hemisphere. Brain 64: 244–272
2. Battersby WS, Bender MB, Pollack M, Kahn RL (1956) Unilateral "spatial agnosia" ("inattention"). Brain 79: 68–93
3. Costa LD, Vaughan Jr HG, Horwitz M, Ritter W (1969) Patterns of behavioral deficit associated with visual spatial neglect. Cortex 5: 242–263
4. Critchley M (1953) The parietal lobes. Arnold, London, p 480
5. De Renzi E, Faglioni P, Scotti G (1970) Hemispheric contribution to exploration of space through the visual and tactile modality. Cortex 6: 191–203
6. De Renzi E, Scotti G, Spinnler H (1969) Perceptual and asso-

ciative disorders of visual recognition: Relationship to the side of the cerebral lesion. Neurology 19: 634–642
7. De Renzi E, Spinnler H (1967) Impaired performance on color tasks in patients with hemispheric damage. Cortex 3: 194–216
8. Faust C (1955) Die zerebralen Herdstörungen bei Hinterhaupts-verletzungen und ihre Beurteilung. G Thieme, Stuttgart, p 111
9. Franceschetti A, François J, Babel J (1963) Les hérédodégénérescences chorio-rétiniennes, vol 1. Masson et Cie, Paris, pp 75–79
10. Geschwind N (1965) Disconnection syndromes in animals and man. Brain 88: 237–294, 585–644
11. Hécaen H (1972 a) Introduction à la neuropsychologie. Larousse, Paris, p 328
12. Hécaen H (1972 b) Neuropsychologie de la perception visuelle. Masson et Cie, Paris, p 320
13. Hécaen H, Angelergues R (1963) La Cécité psychique. Masson et Cie, Paris
14. Hécaen H, Angelergues R (1965) Neuropsychologie clinique des lobes occipitaux. Rapports du 8e Congrès International de Neurologie, III, Vienne, pp 33–42
15. Hécaen H, Penfield W, Bertrand C, Malmo R (1956) The syndrome of apractognosia due to lesions of the minor cerebral hemisphere. Arch Neurol Psychiatr 75: 400–434
16. Kleist K (1934) Gehirnpathologie, 1408. J Barth, Leipzig
17. Lashley KS (1929) Brain mechanisms and intelligence. Univ of Chicago Press, Chicago
18. Leicester J, Sidman M, Stoddard LT, Mohr JP (1969) Some determinants of visual neglect. J Neurol Neurosurg Psychiatry 32: 580–587
19. Lenneberg EH (1961) Color naming, color recognition, color discrimination: a reappraisal. Percept a Motor Skills 12: 275–382
20. Lhermitte F (1972) Les troubles de la vision des couleurs dans les lésions postérieures du cerveau. In: Hécaen H (ed) Neuropsychologie de la perception visuelle. Masson, Paris, pp 241–250
21. Lhermitte F, Chain F, Aron D, Leblanc M, Souty O (1969) Les troubles de la vision des couleurs dans les lésions postérieures du cerveau. Rev Neurol 121: 5–29
22. Luria SL (1967) Color-name as a function of stimulus-intensity and duration. Amer J Psychol 80: 14–27
23. Luria AR (1973) The working brain. Penguin Books, New York
24. Oxbury JM, Oxbury M, Humphrey NK (1969) Varieties of colour anomia. Brain 92: 847–860

Correspondence: Prof. Dr. G. Foroglou, Neurosurgical Clinic of the University, AHEPA General Hospital, Thessaloniki, Greece.

Acta Neurochirurgica, Suppl. 44, 93 (1988)

Concluding Remarks to Preceding Contributions of N. Sichez-Auclair and A. Violon

N. Brooks

University of Glasgow, Department of Psychological Medicine, Glasgow, U.K.

"The contributions of Mrs. Violon and Mrs. Sichez are helpful and stimulating. I entirely agree with the point raised by Mrs. Sichez that post-traumatic personality changes do not come at random. As she correctly points out, head injured individuals form a particular sample of the population, in that they often have had more than one injury, they are young, male, and are often risk takers. Indeed, in our own study of the 5 year outcome after brain injury, 12 of our 42 patients (31%) had some involvement with the police for offences ranging from attempted murder (1 case) to public order disturbances (9 cases). In very many of these cases there had been similar troubles even before the injury.

A further useful feature of Mrs. Sichez's comments is her attempt to identify an underlying mechanism of the changes in personality. She attempts to do this both in terms of psychological mechanisms (is there a change in introversion/extraversion; inhibition/disinhibition, etc?), and pathophysiological or neuropharmacological mechanisms (patterns of lesions; changes in brain chemistry, etc.). With the ever increasing sophistication of brain imaging, such attempts will become more and more realistic, and the opportunity to image brain metabolism and transmitter receptor distributions opens a fruitful area for investigation of the changes in personality.

A problem raised both by Mrs. Violon and Mrs. Sichez is that of nomenclature. Words such as "depression", "paranoia", are used loosely in describing changes after head injury, and whereas depressive or paranoid *symptomatology* may be common, depressive or paranoid *states* which would conform to classic psychiatric terminology are not common. Indeed, Mrs. Violon makes this absolutely clear in her stimulating and precise comments.

Correspondence: Prof. N. Brooks, Ph.D., University of Glasgow, Department of Psychological Medicine, 6 Whittinghame Gardens, Great Western Road, Glasgow, G12 OAA, U.K.

V. Epilepsy and Personality

Acta Neurochirurgica, Suppl. 44, 97 (1988)

Introduction

E. Thiery

Head of the Department of Physiological Psychology and Neuropsychology, State University of Ghent, Belgium

The patients with epilepsy need help and attention from many sides. Neurological and neurosurgical diagnosis and treatment are essential as is the psychiatric and (neuro-)psychological approach to all possible behavioural problems and especially personality problems.

Our eminent colleague L. Sorel would have stressed this if he had been able to come to this convention of our Academia. It is also the idea that stimulated A. Dupuis, a famous Belgian painter, to make an impressive triptych for the Lennox Centre at Ottignies. The left panel pictures show some patients suffering from a scattered mind. The right panel shows the human brain and the work of neurology and neurosurgery. The epileptic patients stands in between. He is represented in the middle panel. His illness threatens him but he has to stay in the world by his and our efforts.

The problem *epilepsy and personality* is a debate which has continued for many years. It is also a very important one, sociologically speaking, as the incidence of this disease is high. Only a small proportion of patients with epilepsy suffer from personality difficulties, but when these occur they frequently interfere with family, social and professional life. Which aspects of the disease will or can influence personality (and language) and at what extent? These and other questions are important topics of actual research and important topics of discussion at the Academia.

Acta Neurochirurgica, Suppl. 44, 98–101 (1988)

Personality Disorders and Epilepsy

M. R. Trimble

National Hospital for Nervous Diseases, Queen Square, London, U.K.

Summary

The relationship between epilepsy and personality disorder is viewed. The historical links are discussed, and it is pointed out that the position has moved from that of assuming that everybody with epilepsy will undergo personality change, to noting in particular changes that may be associated with temporal lobe, especially limbic system, abnormalities. The Geschwind syndrome is briefly described, and the association between aggression and epilepsy explored. It is concluded that some aspects of personality may improve following temporal lobectomy, especially if seizures also improve.

Keywords: Epilepsy; aggression; personality disorders.

Introduction

Although the relationship between epilepsy and psychiatry has been discussed by physicians since the time of Hippocrates, it became a subject of scientific discussion in the 18th and 19th centuries. The modern era may be said to begin with the introduction of the electroencephalogram into clinical practice in the 1940s.

Hughlings Jackson's contributions to epilepsy have been widely acknowledged, and one of the most important has been the clarification of the concept of partial seizures. Thus, prior to the writings of Jackson, epilepsy tended to suggest a generalized convulsive disorder, mainly of motor function. Although sensory phenomena had been described, and the concept of the aura was well known, it was Jackson who more fully elaborated the idea of partial seizures, the motor form ultimately being designated "Jacksonian" by the French neurologist Charcot. With regards to the psychiatric aspects of epilepsy, Jackson described "the uncinate group of fits", and noted patients who presented with psychic changes as a manifestation of the seizure. He thus described a forerunner of what later became referred to as temporal lobe epilepsy with associated psychomotor seizures, identified clearly with the electroencephalogram. In our present classifications, the terms simple and complex partial seizures describe the kind of attacks which Jackson delineated.

In order to understand our present thinking with regards to the relationship between personality disorders and epilepsy, it is valuable to consider the divisions of historical thinking on the subject as outlined by Guerrant *et al.* (1962). These authors identified four periods as shown in Table 1. The first, referred to as the period of epileptic deterioration, which approximated to the end of the last century, assumed the position that because patients had epilepsy, this was enough for some deterioration of their personality and intellect to occur. These ideas stem directly from the degeneracy theory of mental disorder prevalent in Continental Europe at that time, in which it was considered that neuropsychiatric illness, including epilepsy, was the result of a progressive hereditary degenerative condition which would pass from one generation to the other and ultimately lead to imbecility.

The period of epileptic character, referring to concepts which were prevalent in the early part of this century, lead to a marked change of thinking. The epilepsy itself, and any associated behaviour or personality changes, were seen as secondary to some underlying third principle, which was common to both. One of the uniting factors was the constitution of the patients, and their "personality". This line of thought was pursued by Pierce Clark (1929), and it was claimed

Table 1. *Changing Ideas of the Relationship Between Epilepsy and Psychopathology*

Period of epileptic deterioration	(–1900)
Period of the epileptic character	(1900–1930)
Period of normality	(1930–)
Period of psychomotor peculiarity	(1930–)

after Guerrant et al. (1962)

that it was possible to discern certain patterns of developmental personality traits which could be identified in patients prior to the onset of seizures, and implied an epileptic constitution. The whole of this era became bound up with Freudian psychodynamics such that epilepsy, as with a number of other conditions, was thought to reflect infantile motives and inadequate development of individuals' affects and instincts leading to the later development of symptoms. From these ideas arose a number of concepts in psychomatic medicine which related to various disease "personalities", and in the case of epilepsy to the "epileptoid" character and to epileptic equivalents.

In the mid-part of the century, views changed such that it was considered that patients with epilepsy were perfectly normal, and while some may undergo changes of personality secondary to head injuries, to the long-term prescription of anticonvulsant drugs, or to the consequences of psychosocial stigmatization, the very fact that patients had seizures did not leave them susceptible to personality changes. This "era of normality" was most forcefully stated by Lennox (1944), and is still a prevalent view of many researchers today. However, an alternative view, which to some extent is a return to the "period of the epileptic character", is that not all patients with epilepsy are liable to undergo personality changes, but it is those with a chronic temporal lobe abnormality, and limbic epilepsy, who are most susceptible. This "period of psychomotor peculiarity" was most strongly stated by Gibbs and Stamps (1958) thus: "the patient's emotional reactions to his seizures, to his family and to his social situation are less important determinants of psychiatric disorder than the site and type of the epileptic discharge".

Epilepsy and Personality Disorder

In more recent investigations, it is usual to try to assess personality using standardized rating scales, and a number of studies have been carried out trying to define differences between patients with temporal lobe and other forms of epilepsy. On the whole, many of these studies have been negative (Trimble and Perez 1982), although there is more agreement between authors that patients with epilepsy when compared to other patients who do not have epilepsy do indeed show more psychopathology. However, the interpretation of this would be compatible with a number of different view points, but the central argument relates to differences within epilepsy populations between those who have temporal lobe lesions and others.

Some studies clarify this issue. Bear and Fedio (1977) developed their own rating scale of 18 behavioural features drawn from the literature which were supposedly associated with temporal lobe epilepsy. They gave this to a group of patients with temporal lobe epilepsy and compared their profiles to controls with neuromuscular disease and to healthy counterparts. Their scale appeared to separate out the temporal lobe group, in particular emphasizing certain traits such as obsessionality, dependency, and religious and philosophical concerns.

Although these data have been criticised, they have been partially replicated by other groups (Hermann and Riel 1981). In a more recent study, Nielson and Kristensen (1981) used this same scale in patients with temporal lobe epilepsy, but compared a group with a mediobasal focus to those with a lateral neocortical focus of their abnormality as recorded on the electroencephalogram. The former were shown to have more personality changes, including such traits as hypergraphia, elation, guilt and paranoia, compared to those with a lateral focus. Using the Minnesota Multiphasic Personality Inventory (MMPI), Hermann *et al.* (1982) examined patients with temporal lobe epilepsy, but divided them into different groups depending on the substance of their auras. Those with an aura or fear scored significantly higher on several items of psychopathology compared with those with other auras, or to patients with generalized seizures. These two studies emphasize an important point. Thus, they suggest that temporal lobe epilepsy does not represent a uniform condition, and that there is a group of patients, with mediobasal lesions, and possibly who have auras of fear, who are more susceptible to develop psychopathology. Since an aura of fear, and the mediobasal focus imply limbic system lesions, it is suggested that it is patients with "limbic epilepsy" who may be susceptible to the development of psychopathology, because they have a chronic disturbance of their limbic system function.

The Geschwind Syndrome

The concept of personality change secondary to chronic temporal lobe lesions has been most strongly made by Geschwind and his colleagues (Waxman and Geschwind 1975, Geschwind 1979). They emphasize in particular hyposexuality, religiosity and hypergraphia—a tendency towards extensive and compulsive writing. The hypergraphia of these patient is particularly interesting, as is the religiosity. The hypergraphia is characteristically meticulous, containing moral or religious overtones, and there is preoccupation with detail and often a compulsive quality to much that is written. Repetition of words or sentences may be seen and variants including the hiring of public stenographers to record for patients, or extensive drawing, or even painting. The religiosity may take the form of sudden religious conversions, or of an increasing preoccupation with religious behaviour, e.g. compulsive Bible reading and church attendances, and philosophical notions about the nature of seizures, why the patients have them, and on occasions a religious explanation as to why the patients should have been "chosen" to have epilepsy.

It is quite clear that many patients with temporal lobe epilepsy do not show this constellation of symptoms, and it is not universally found. Further, some aspects of the syndrome are easily missed unless specifically asked for. For example, it is not customary to ask patients if they keep extensive diaries, write books, or are overconcerned with philosophical matters. Nonetheless, when carefully probed, the Geschwind syndrome is commoner than otherwise thought. Further, it may fluctuate in its intensity, and should not necessarily be thought of as being counterproductive to the individual's life in society.

Bear (1986) discusses three different aspects of behaviour which comprise the interictal behaviour syndrome. The first relates to alteration of physiological drives, including sexuality, aggression and fear. The second group he summarizes as "nascent intellectual interests", showing preoccupation with religious, moral, cosmological or philosophical themes. The third group are termed "dispositions" in association with interpersonal relationships. Under this heading he includes preoccupation with details (obsessiveness), circumstantially, and a tendency to prolong and deepen social encounters, sometimes referred to as viscosity. He contrasts these aspects of the behaviour of temporal lobe epilepsy with the Kluver-Bucy syndrome where aggression is reduced, sexual behaviour is increased, and social cohesion is diminished. The possibility that the interictal syndrome represents abnormal sensory-limbic connections, possibly through kindling, is suggested, this increased limbic permeability being contrasted with the limbic agnosia of the Kluver-Bucy syndrome.

Aggression and Epilepsy

The possibility that patients with epilepsy are prone to aggressive outbursts has attracted considerable attention but there are very few studies. A major problem has been the difficulty of recording and quantifying aggressive behaviour, particularly as it is almost always interpersonal and situational in nature.

During automatisms, aggressive behaviour is sometimes seen, but it is poorly directed and often provoked by inadequate handling of the confused patients. Increased aggressive behaviour as a personality trait in epileptic patients has however been noted in several studies, although this is still an area of considerable controversy. For example, Kligman and Goldberg (1975) examined eight controlled studies where aggression in epilepsy had been investigated, and noted only two that supported the relationship, although they severely criticised the methodology in all of them. Ictal aggression is occasionally observed and has forensic relevance in that the violent episodes may lead to harm of a third person (Fenwick 1986).

Falconer (1973) noted a high prevalence of aggression in patients with difficult to control epilepsy who were assessed for temporal lobectomy. Certainly following temporal lobectomy, of all of the behavioural abnormalities described in patients with epilepsy, it is the aggression that is most likely to resolve. Serafetinides (1965) studied 100 patients operated on for temporal lobectomy and noted that a history of explicit aggressive outbursts was associated with an early onset of seizures, with left-sided resections, and with epigastric, autonomic and complex automatisms during seizures. Following operation aggression improved, but only in those patients whose epilepsy also improved. Several authors have suggested that aggressive behaviour may relate to abnormal limbic system, especially amygdala, activity, which is in keeping with considerable evidence from the animal literature that the amygdala and hypothalamus in particular are involved in the modulation of aggressive responses.

Conclusions

The association between epilepsy and personality disorder has been under investigation for a number of years and some tentative conclusions may be drawn. It is suggested that patients with limbic system dysfunction are the most susceptible to undergo changes of personality, and that the latter reflect a chronic ongoing organic personality change. It does not necessarily have to be irreversible, and some features of the personality changes, for example aggressive behaviour, seem ameliorated by procedures such as temporal lobectomy particularly if seizures are also brought under control. Many who deny the existence of such personality changes often fail to seek in the clinic some of the behavioural associations which have been discussed in this paper, while those who support the concept of the syndrome believe it may be more commonly encountered if appropriate areas of the patient's life are explored. The personality changes themselves need not necessarily be viewed pejoratively, and reflect upon a number of traits which have great importance for human aesthetic and social behaviour. However, if they do impede a patient's progress, then further exploration of them would seem worthwhile.

References

1. Bear DM, Fedio P (1977) Quantitative analysis of interictal behaviour in temporal lobe epilepsy. Arch Neurol 34: 454–467
2. Bear DM (1986) Behavioural changes in temporal lobe epilepsy: conflict, confusion, challenge. In: Trimble MR, Bolwig TG (eds) Aspects of epilepsy and psychiatry. John Wiley & Sons, Chichester, pp 19–29
3. Falconer M (1973) Reversibility by temporal lobe section of the behavioural abnormalities of temporal lobe epilepsy. New England Journal of Medicine 289: 451–455
4. Fenwick P (1986) Aggression and epilepsy. In: Trimble MR, Bolwig TG (eds) Aspects of epilepsy and psychiatry. John Wiley & Sons, Chichester, pp 31–57
5. Geschwind N (1979) Behavioural changes in temporal lobe epilepsy. Psychological Medicine 9: 217–219
6. Gibbs FA, Stamps FW (1958) Epilepsy handbook. Charles C Thomas, Springfield Illinois
7. Guerrant J, Andersen C, Fischer A, Weinstein MR, Jarros RM, Deskins A (1962) Personality and epilepsy. Thomas, Springfield
8. Hermann BP, Reil R (1981) Interictal personality and behavioural traits in temporal lobe and generalized epilepsy. Cortex 17: 125–128
9. Hermann BP, Dickmen S, Schwartz MS, Karnes WE (1982) Interictal psychopathology in patients with ictal fear: a quantitative investigation. Neurology 32: 7–11
10. Kligman D, Goldberg DA (1975) Temporal lobe epilepsy and aggression. J Nerv Ment Dis 160: 324–341
11. Lennox WG (1944) Epilepsy. In: Hunt. Handbook of personality and behaviour problems. Ronald Press, New York
12. Nielsen H, Kristensen O (1981) Personality correlates of sphinoidal EEG foci in temporal lobe epilepsy. Acta Neurol Scand 64: 289–300
13. Pierce-Clark L (1929) A psychological interpretation of essential epilepsy. Brain 43: 38–49
14. Serafetinides EA (1965) Aggressiveness in temporal lobe epilepsies and its relation to cerebral dysfunction and environmental factors. Epilepsia 6: 33–42
15. Trimble MR, Perez MM (1982) The phenomenology of the chronic psychoses of epilepsy. In: Koella WP, Trimble MR (eds) Temporal lobe epilepsy, mania, schizophrenia and the limbic system. Karger, Basel, pp 98–105
16. Waxman SG, Geschwind N (1975) The interictal behaviour syndrome of temporal lobe epilepsy. Arch Gen Psychiatry 32: 1580–1586

Correspondence: M. R. Trimble, FRCP, FRC Psych, Consultant Physician in Psychological Medicine, National Hospital for Nervous Diseases, Queen Square, London WC1N 3B6, U.K.

Acta Neurochirurgica, Suppl. 44, 102–105 (1988)

Surgical Treatment of Epilepsy. Restoration of Personality?

F. J. Gillingham

Edinburgh, U.K.

Summary

352 epileptic patients were studied by basic science, clinical neurological and psychological teams to assess the effects of epilepsy and its treatment on personality. Those who did not response to well-tried medication were subjected to surgical treatment and short- and long-term follow-up of the results of treatment carried out. The changes in personality following operation and their relationship to the reduction of seizures were recorded. Temporal lobe epilepsy is closely related to personality disorder and the benefits of classical temporal lobectomy on the reduction of seizures and restoration of personality is confirmed. However it is also confirmed that for the best results the earliest referal is necessary *i.e.* as soon as the failure to respond to adequate therapy is evident. This equally applies to amygdalotomy for violent and destructive behaviour associated with epilepsy particularly in children and also to stereotactic "central" lesions for non-focal intractable epilepsy. The "prefered" pathway for the propagation of the epileptic discharge and its relationship to neurochemical changes occurring during the increased neuronal activity associated with focal motor seizures is described.

Keywords: Epilepsy—focal/G.M.; behavioural changes—psychopathic personality; modification/medical; surgical therapies; long and short term follow-up; pattern of epileptic discharge—preferred pathways; neuro-chemical changes.

As we have heard today and read over the years the definition of personality and its measurement remains difficult. In addition many publications have not included controls or matched groups. Even then some have shown little difference between the personality of the epileptic and that of the normal population. As regards surgical treatment and the assessment of results, selection for consideration of operation, first by the family doctor, then the general physician, neurologist and finally the surgeon together with the necessity for long follow-up makes the task complex and perhaps less reliable than we would wish. Also there are extremes of personality change, from the violent and destructive, better known as behaviour disorder to the mildly restless, irritable, unpredictable, depressed and anxious some of which may well be the outcome of shame and fear, so-called personality disorder. The variables are bewildering and a long follow-up is essential.

During the five year period 1967–1972 we set up a research team to study certain aspects of epilepsy and its surgical management in 352 patients referred to the University department of Edinburgh. It included medical and surgical neurology, neurophysiology, neurochemistry and pharmacology, neuropathology and neuro-radiology but not of course modern scanning which began in 1971. We experimented with continuous video and simultaneous EEG recordings and chronically implanted electrodes. Both were discontinued, the first because at that stage of development the information was insufficient and costly in staff-time, especially nursing. The second because of ethical worries and some concern about infection and haemorrhage. Telemetering now seems to be the solution for the more difficult problems. However serial interictal EEG studies, even over a period of up to six months, sphenoidal electrodes and corticography were routinely used. Stereotaxy was at its height and for precise target siting we used depth microelectrode recording and audio-monitoring (Gaze *et al.* 1964).

Personality changes were investigated by a neuropsychiatrist, two medical psychologists, rehabilitation and resettlement staff including speech-therapists, social workers and the hospital priest. Reports from this last group were published by Cairns, Naughton and Hitchcock (1973, 1974). Related work by other members of the team have been published over the years by Ashcroft, Donaldson, Gaze, Harris, Hitchcock and myself.

The purpose of this paper is to discover what happened to the 69 operated patients of the 352 studied by our medical psychologists. All 69 suffered from medical refractory epilepsy and some degree of personality disorder. This is a relatively high percentage for operation compared to some other series but it does illustrate the considerable problem of screening for possible operative treatment. 16 of the 69 were lost to follow-up after one to two years and two died, one of asphyxia in her bath two years after operation and one of a diffuse astrocytoma. One patient became intellectually and neurologically disabled after a cerebral vascular accident. Thus there were 50 patients with a follow-up of between 5 and 20 years—an average of 10,8. Of these 50, seven were excluded from the final study because of structural lesions or irrelevant operations such as shunts. So finally there 43 specific operations for epilepsy and personality disorders.

We have little experience of hemipherectomy and none of callosal section even now. Therefore I propose to confine my remarks to the 43 patients treated by temporal lobectomy (11 patients), partieto-occipital cortical excision (2 patients), stereotactic amygdalotomy (16 patients), and subcortical "central" lesions (14 patients).

But first we should review briefly the *short-term follow-up* of personality in the 352 patients studied by Cairns and Naughton (1974). They used standard psychometric techniques—questionnaires, adaptive behaviour scales and experimental situations of a mildly stressful nature. They endeavoured to use matched comparisons for the various groups *e.g.* age, sex, operative, non-operative, etc. Their conclusions were that observational techniques provided the most useful data for the assessment of results of surgical treatment *i.e.* adaptive behaviour scales and experimental situations. With a follow-up of not less than six months in those subjected to surgery they found varying degrees of postoperative improvement in personality or behaviour in all of our four groups, some markedly so, which in general ran parallel to a decrease in severity and frequency of seizures. In their previous paper (1973) they had also compared cognitive function of 106 being considered for operative treatment. Their findings confirmed the studies of Pond (1960) and of Falconer (1970) that the early onset of temporal lobe epilepsy is likely to be associated with poor cognitive test performance. In turn this is associated with behavioural disturbance the difficulty of management of which may lead to earlier referal for surgical treatment. Also when there is relief or remission of seizures there is often improvement in psychological function. This was our psychologists' conclusion regarding the short-term results of surgery. It is also important to note that in the group studied by Cairns and Naughton only those with clear-cut temporal lobe epilepsy had significant behaviour problems as opposed to those of personality, the extreme of which is violence and destructiveness. They also noticed a greater improvement of cognitive function especially in rote verbal memory (in those under 20) in those specifically referred for possible surgery an opposed to those not referred for surgery—suggesting perhaps that only those who have deteriorated should be operated upon *i.e.* using surgery as a last resort.

This study and others suggest then that the more severe behavioural and personality changes are more specific to temporal lobe epilepsy than other types of seizures if they are not controlled early. Also if other types of seizures are not controlled adequately by medication or surgery or both, deterioration of cognitive function and the related personality traits become evident. This supports the hypothesis that there is a continuing interictal epileptiform process interfering with the psychological functions of "attention and registration" indicative of involvement of the activating system in epilepsy in general and of the limbic system for temporal lobe epilepsy in particular. These points may gain more emphasis when we consider the long-term view of the particular operative groups.

The type of operation selected was related to the type of epilepsy and behaviour repercussion *e.g.* the predominance of violence and destructiveness over seizures indicated the need for ablation of the amygdala. A clear cut unilateral EEG focus and clinical evidence of temporal lobe seizures pointed to a classical temporal lobectomy. Focal seizures but an EEG focus at a different site with serial studies suggested that exploration, corticography and possible excision was the best approach. Finally those patients with medically intractable, frequent and severe grand mal seizures were treated by stereotactic "central" lesions. These last tended to be in a somewhat older age group whereas the majority of the others were young adults or children, the youngest of whom was six years old.

In the *short-term postoperative study*, slight but definite improvement in cognitive function was observed after ablation of the amygdala and after the cortical operations, even on the dominant side. However this was not found after the "central" lesions but the lack of improvement may have related to late referal and the older age of this group of patients. It is worthy of note that all 14 of this group were being considered preoperatively for prolonged institutional care because of uncontrolled seizures, one having five major fits a week.

Now let us look at the *long-term follow-up*. A simplified adaptive behaviour scale similar to that employed by Gibson (1974) was used and diaries recording seizures were kept by patients and relatives reasonably well.

Two patients with right-sided temporo-parietal cortical operations were related to previous head injury. One, a female of 19 required scar and some cortical excision after corticography. With two focal or grand mal fits a month and minimal personality problems she has done well, is married and has two children. She remains on minimal medication and has only a rare attack of a focal character. The other, a male of 17, dull mentally and with frequent grand mal and focal fits was treated similarly. Now he has no more than one grand mal attack a year on continuing medication. He lives independently and is occupied in simple work on the land. His interpersonal relationships are good and he has a good work record. Fortunately both these patients had caring relatives and their rehabilitation and resettlement were safeguarded at hospital out-patients and at home.

In the temporal lobe epilepsy group there were 11 with clear-cut unilateral seizures. Two children were violent, destructive and had frequent attacks. One of them, a girl of 11 and the most destructive responded dramatically in respect of behaviour becoming an affectionate quiet child. Seizures ceased but she was continued on reduced medication. The other, a boy of 12 relapsed after a good first year in respect of behaviour but seizures were markedly reduced in frequency. In his case the EEG studies in retrospect were less clear-cut and he had been referred late. Of the other 9, all adults, 2 had epilepsy as the main problem and these dramatically lessened to rare attacks following operation. In both interpersonal relationships remained good. Of the 2 the most successful is a practicing lawyer in the city although he required further excision of the mesial structures one year after lobectomy because of incomplete reduction of fits. 4 of the 7 remaining were greatly improved as regards control of seizures and behaviour. Both had been seriously disabled. In 3 results were poor, the worst a female of 35 whose fits were reasonably controlled but the disturbance of behaviour was unchanged. Grossly unstable and suicidal the still spends much of her time in a psychiatric hospital. Histology was positive in 5 of the 11 patients. 2 showed a concentration of gliosis in the mesial structures and in 3 there were small scattered areas of gliosis throughout the temporal lobe. This more widespread change occurred in patients who did less well and they would now qualify for more intensive investigation using the sophisticated techniques of Rassmussen (1980) and Talairach (1980).

The results of operation in this small group, namely between 50–60% improvement approximate to many larger series with shorter follow-up using a similar less highly sophisticated approach. This is important to note in view of the relative paucity of resources in the developing countries in which the incidence of remediable epilepsy is certainly no less.

Amygdalotomy was carried out in 16 patients all with moderately severe or severe behavioural disorders associated with seizures which were often not precisely those of a temporal lobe pattern and less commanding as a problem. 5 were children, the youngest six years of age. The rest were young or older adults, the oldest forty-one. 8 were severely disturbed and required bilateral procedures before behaviour was modified. All were improved in the original short-term studies. 9 of the 12 patients were considerably improved in respect of control of their behaviour and 7 less so. Children benefitted the most. Seizures were better controlled in 8 of the 16 but when they were it was regarded as a bonus. All required continued medication. Over the years there was a fall-off of approximately 20% in maintained improvement of behaviour after a unilateral operation and also in 2 patients who had bilateral procedures. The worst result was in the adult of 41 years. It is perhaps remarkable that in the group as a whole as reported by Cairns and Naughton (1973, 1974) and by Vaernet (1974) there is so little psychological impairment from these relatively large bilateral lesions.

Our final group of "central" or subcortical stereotactic lesions were carried out for intractable grand mal or non-focal seizures which were so severe before operation that the patients were under consideration for institutional care. My interest in this field began in 1958 when a male of 47 was referred for control of severe unilateral tremor. Tremor was abolished by a relatively small lesion of the globus pallidus adjacent to the internal capsule. The interest in this patient was the coexistence of disabling attacks of obsessive calculation occurring fifteen times a day during which he would be withdrawn for several mintues—a condition similar to that described by Penfield as "forced thinking" and considered by him to be a form of epilepsy. These curious attacks were virtually abolished *i.e.* reduced to one or two a week for the full length of follow-up. Thus we were alerted to the possibility of the existence of a "preferred" pathway for the propagation of the epileptic discharge (Gillingham 1968). It was the work of Jinnai (1963) on lesions in the field of forel which led to further explorations of this observation. Our 14 patients were all adults, the youngest being a female of twenty-two years. 13 showed dull personalities preoperatively but without significant behavioural difficulties. They were all on prolonged heavy medication without benefit. 3 were occasionally unpredictable and irresponsible and 1 was mentally defective. A medium-term study of these patients was presented in 1980 (Gillingham 1980). Of the group 1 has died and 1 seriously disabled from a cerebrovascular accident. Further follow-up of the remaining 12 has shown a maintained reduction of 50% in the frequency and severity of their fits. In none were they abolished. Medication was maintained but in 3 it was reduced. The 3 with behavioural difficulties were improved. None are institutionalized and half of them are working, one in protected workshops. The others are cared for by their families. One, the youngest died in her bath two years after operation and up to the time of her death the best result. 8 patients had bilateral procedures and they carried some morbidity in six which was fortunately short-lived. Three sites of operation were employed, the field of Forel, the ventrolateral nucleus of the thalamus and the pallidum. It is important to note that the best results in this small series came from lesions of the pallidum adjacent to the internal capsule supporting the original observation of 1958. It is also supported by the work of Caveness (1980) on the utilization of glucose by various structures within the basal ganglia during induced seizures in the monkey. He found that in the lateral pallidum it was twice that of the other structures studied including the ventrolateral nucleus of the thalamus, the VPM of the thalamus, cerebellar cortex and medial pallidum. He felt that these results provided fresh insight into the location and extent of increased neuronal activity in focal motor seizures and that they might provide a target for stereotactic therapeutic procedures.

References

1. Ashcroft GW, Dow RC, Emson PC, Harris P, Ingleby J, Joseph MH, McQueen JK (1974) A collaborative study of cobalt lesions in the rats a model for epilepsy. In: Harris P, Mawdsley C (eds) Epilepsy. Proc of the Hans Berger Centenary Symp, Churchill Livingstone, pp 115–124
2. Cairns VM (1974) Epilepsy, personality and behaviour. In: Harris P, Mawdsley C (eds) Epilepsy. Proc of the Hans Berger Centenary Symp, Churchill Livingstone, pp 256–267
3. Cairns VM, Naughton JAL (1973) Surgical treatment of epilepsy. Evaluation of psychological effects. In: Parsonage MJ (ed) International Bureau on Epilepsy. Prevention of epilepsy and its consequences. London
4. Caveness W, Kato M, Hosokawa S, Malamut B, Wakisaka S, O'Neil R (1980) Propagation of focal motor seizures in the monkey. In: Wada JA, Penry JK (eds) 10th Epilepsy Int. Symp., p 492
5. Falconer MA (1970) Significance of surgery for temporal lobe epilepsy in childhood and adolescence. J Neurosurg 33: 233–252
6. Gillingham FJ, Watson WS, Chung S, Yates C (1980) Central brain lesions for the control of intractable epilepsy. In: Wada JA, Denry JK (eds) 10th Epilepsy Int. Symp., pp 251–255
7. Gibson RM, Maxwell RDH (1974) The effect of temporal lobectomy upon severe disturbances of behaviour. In: Harris P, Mawdsley (eds) Epilepsy. Proc of the Hans Berger Centenary, Churchill Livingstone, pp 272–275
8. Hitchcock ER (1973) Observations on the development of an assessment scheme for amygdalotomy: surgical approaches in psychiatry. E & S Livingstone, Edinburgh
9. Jinnai S, Nishimoto A (1963) Stereotactic destruction of Forel-H for treatment of epilepsy. Neurochirurgia 6: 164–176
10. Pond DA, Bridwell BH (1960) A survey of epilepsy in fourteen general practices. Epilepsia 1: 285–299
11. Rassmussen T (1980) Surgical aspects of temporal lobe epilepsy. Acta Neurochir (Wien) [Suppl] 30: 13–24
12. Talairach J, Szikla G (1980) Application of stereotactic concepts to the surgery of epilepsy. Acta Neurochir (Wien) [Suppl] 30: 35–54
13. Vaernet K (1974) The relationship of psychomotor epilepsy to pathological aggression. The effect of amygdalotomy. In: Harris P, Mawdsley C (eds) Epilepsy. Proc of the Hans Berger Centenary Symp. Churchill, Livingstone, pp 222–226

Correspondence: F. J. Gillingham, M.D., Prof. emeritus of Neurosurgery, Easter Park House, Barnton Avenue, Edinburgh EH4 6JR, U.K.

Acta Neurochirurgica, Suppl. 44, 106–110 (1988)

Epilepsy and Verbal Behaviour

Y. Lebrun

Neurolinguistics Department, School of Medicine, V.U.B., Brussels, Belgium

Summary

A review is given on the various ways of influence between paroxysmal brain activity and verbal behaviour.

In some cases the performance of specific linguistic tasks may precipitate an epileptic fit. Often epileptic discharges interfere with speaking, writing and/or understanding of speech or may result in palilalias, verbal automatisms or other forms of disturbance of verbal performance. A different but not uncommon aspect of epilepsy is hypergraphia. In some few cases even a speech deblocking effect of epileptic discharges has been observed.

Keywords: Epilepsy; language; language-induced epilepsy; ictal verbal behaviour; hypergraphia.

Introduction

Even a cursory glance through the voluminous literature on epilepsy will easily convince one that the verbal behaviour of epileptics may be affected in many different ways. The relationships between language and epilepsy are both complex and bi-directional. On the one hand, linguistic activity may elicit paroxysmal brain discharges. On the other hand, epileptic fits may variously interfere with linguistic activity. The purpose of this presentation is to sketch briefly some features of the reciprocal relation between language and epilepsy.

Language-Induced Epilepsy

In a number of patients the performance of some linguistic task may readily precipitate an epileptic fit. The language-induced attack may be clinical or remain subclinical. When it is clinical, its manifestations may be so subtle as to be perceptible to the sufferer only. Conversely, the ictal episode may begin as a partial seizure and evolve into a generalized attack if the precipitating verbal activity is not discontinued. Indeed, the incitant may at times directly bring on a grand mal crisis.

Language-induced epilepsy is said to be primary when only verbal stimuli have been observed to precipitate convulsive spells. It is said to be secondary when, in addition to verbal, some nonverbal stimuli are known to be effective in bringing about epileptic fits. It is to be noted that patients with primary language-induced epilepsy may have seizures, including grand mal convulsions, provoked by unknown stimuli.

The epileptogenic power does not seem to be the same for all linguistic activities. To judge by the number of cases in the literature, reading is much more liable than any other verbal tasks to elicit paroxysmal brain discharges. The most frequently reported clinical manifestation of *reading epilepsy* is jaw jerking. In some cases, the symptom is felt by the patient but is not perceptible to an observer. Jaw jerking may be accompanied by a feeling of tightness in the throat or of pressure in the head. In a few cases, however, myoclonias different from jaw jerking have been noted. For instance, a patient of Alajouanine *et al.* (1959) evidenced convulsions of both upper extremities under the influence of reading. In some cases, epileptic fits are more likely to occur when the patient reads aloud than when he reads silently. In other cases, both activities are equally liable to bring about ictal episodes. It is still uncertain which features of the reading process trigger off the paroxysmal brain activity. Some researchers consider the eye movements, especially the wide leftward gaze sweep at the end of each line of reading, to be the epileptogenic stimuli. Others, on the contrary, hold the subliminal articulatory movements that accompany reading to be responsible for the seizures.

Not only reading but also writing may precipitate epileptic spells. However, *graphogenic epilepsy* seems to be far less frequent than reading epilepsy.

A few patients have been reported who had *paroxysmal spells caused by the production or the perception of speech*. Only the production of certain kinds of speech or the auditory perception of certain kinds of verbal messages are liable to act as precipitants. For instance, Herskowitz *et al.* (1984) described a two-year-old toddler who frequently had partial motor seizures when singing, or humming or when reciting nursery rhymes. Ordinary conversational use of speech did not induce any epileptic fits. It may therefore be conjectured that it was the rhythmic component of the songs and nursery rhymes that acted as an incitant. As an example of epileptic discharges caused by the perception of speech, a case reported by Forster *et al.* (1969) may be quoted. Their patient had both generalized tonico-clonic attacks with loss of consciousness and partial motor seizures without loss of consciousness. Most of the latter occurred while listening to any of three different male announcers. Presumably the patient's brain was abnormally tuned to some phonetic features shared by these three voices.

Patients have also been described who were sensitive to reading, writing and speaking. Such cases are not numerous. Moreover, in the majority of them epileptic fits could be precipitated by writing or speaking only after the brain had been primed by reading, *i.e.* only after a spell of reading epilepsy had taken place. This seems to confirm the view that of all linguistic activities reading has the greatest epileptogenic power. This greater power may be due to the fact that the perception of a succession of black letters or words on a white page may have something in common with intermittent photic stimulation which is known to be epileptogenic in some individuals. It should be noted, however, that in patients with primary language-induced epilepsy, including primary reading epilepsy, such provocative factors as sleep, hyperventilation as well as intermittent photic, optokinetic and visual pattern stimulation are usually ineffective.

Ictal Verbal Behaviour

In some people, verbal activity precipitates paroxysmal episodes. In others, on the contrary, it is interfered with by seizures. Verbal behaviour may be affected in different ways by epileptic discharges. In other words, there exist various forms of ictal verbal behaviour.

In grand mal seizures consciousness is typically lost and all verbal activity is suspended. However, there may be some involuntary vocal production. For instance, the convulsion may begin with a scream, to so-called initial epileptic cry. When the patient recovers consciousness after the convulsive phase, there may be temporarily an abnormal mental state with confused use of language.

During epileptic spells, the patient may be unable to speak even if he does not lose consciousness completely. He may or may not remain able to understand why is said to him. If speech comprehension is preserved, the temporary loss of oral expression is usually called "speech arrest".

While some patients are stricken with dumbness during seizures, other produce involuntary vocalizations. These may be loud screams. Sometimes, as in a case reported by Wohlfart *et al.* (1961), the shouts alternate with expletives. Following Benedek (1925) involunatry production of screams during seizures is sometimes called "klazomania".

The compulsive vocalizations may take the form of palilalias. Various forms of paroxysmal palilalias have been observed. Some patients repeat a great number of times a speech segment (word, phrase, or sentence) which they were meaningfully using when the seizure started. In other cases, the patient cannot help giving palilalic answers to the questions he is asked during his epileptic spell. Still others keep repeating the same irrelevant word, phrase or sentence throughout the paroxysmal episode. This palilalia may remain invariant from fit to fit. For instance, Falconer (1987) described a Polish polyglot who would repeat "I beg your pardon ... I beg your pardon .., I beg your pardon" through each of his seizures. At the same time he was unresponsive to the outside world. Afterwards he was amnesic for the attack and for his repetitive speech behaviour.

Repetitive utterances produced in paroxysmal states are sometimes called "ictal speech automatisms" or "verbal automatisms". However, the latter phrase is also used to refer to the unconscious and generally disordered performance of a meaningful verbal act during a partial epileptic fit. For instance, Souques (1928) published a fragment of a letter in the middle of which a seizure started. The first part of the letter was written neatly. In the second part, on the contrary, the graphic signs are erratic and the page is smeared. A comparable example was published by van Dongen en Bak (1983). A school-child had an absence while writing. She continued writing but elongated a number of letters. In either case epilepsy caused a disturbance of graphomotricity. The patients formed their letters erratically during the seizure but wrote the appropriate words. It may happen, however, that the patient uses inadequate words. Indeed, he may write gibberish. One of the first descriptions of paroxysmal jargonagraphia seems to be

Hughlings Jackson's report on Dr. Z. (1880, 1898), actually Dr. Arthur Thomas Myers. A note written by this physician while he was having a petit mal crisis consists of pure semantic jargon. The note is unintelligible although all the words it comprises are English words. Sometimes, the patient produces neologisms instead of words. Souques (1928) reproduced a letter the second part of which is replete with non-words in addition to perseverations.

The last two examples show that verbal performance may be severely disturbed during seizures. This paroxysmal aphasia may take different forms, some of which are reminiscent of classical motor aphasia, others of classical sensory aphasia, and still others of global aphasia.

Paroxysmal aphasia is interesting among other things because it can help throw some light on the controversial issue of whether classical sensory aphasics are aware or unaware of their deviant verbal production. Traditionally patients with Wernicke's aphasia, especially patients with jargon aphasia, are considered anosognosic. Some authors, and notably Poeck (1972), have opposed this classical teaching and claimed that sensory aphasics are not unaware of their language disorder. Since epileptics who have sensory aphasia during their seizures usually recover their full linguistic competence when the attack is over, one can ask them how they felt while they were having the seizure. A patient of Alajouanine and Sabouraud (1960), explained that when she used language during a seizure she always knew what she wanted to say, but she did not know whether or not she was wording her thoughts correctly. "Je ne sais pas ce que je dis" (I don't know what I say), she stated. However, when she said something which was meant to be serious and her speech partner started to laugh, she concluded that she had been using wrong words. "Je me rends compte que ça ne doit pas être correct, puisque les gens rient, mais (moi) je ne l'entends pas" (I realize that what I have been saying must be inappropriate, since people laugh, but I myself don't hear my errors). When expressing herself in writing, the patient was equally unable to judge the adequacy of her written production. On the other hand, this woman knew during her seizures that she could not understand what she heard. In other words, even in the middle of an attack she was aware of her comprehension disorder.

Partial awareness of one's language impairment was also present in Lecours and Joanette's case (1980). The patient had paroxysmal aphasia primarily of the sensory type. Throughout his ictal episodes he was conscious of his linguistic impairment. Specifically, he realized that he could not understand language. At times, during spells he would deliberately turn on his wireless or go through his mail to find out whether verbal comprehension was returning, indicating that the fit was drawing to an end. The patient was also aware of the inadequacy of his verbal output during epileptic spells and therefore purposely refrained from talking whenever possible. When he did speak, he sensed that what he was saying was incorrect but he did not know exactly what was deviant in his language. "I know that certain words I say are not correct", he explained, "but I do not know which ones and I do not know how I pronounce them". This Canadian patient was totally unaware of the fact that he would sometimes insert a short English idiom or catch-phrase in his French-like jargon. He did not know, and was very surprised to hear from the investigators, that he had a predilection neologism (tuware), and phonological variants thereof, which kept recurring in his jargon from epileptic fit to epileptic fit. On the other hand, the patient was well aware of the fact that he tended to recover his writing skills somewhat faster than his speaking skills, and he would spontaneously resort to paper and pencil rather than to speech if during the anticlimax of an attack he felt the need to communicate verbally. In summary, during paroxysmal episodes, the patient was conscious of having aphasia without being able to perceive his aphasic errors.

Hypergraphia

Under the influence of epilepsy verbal activity may not only decrease, as in patients with speech arrests or paroxysmal motor aphasia, but also increase. More specifically, a number of epileptics become overproductive in writing. They evidence hypergraphia. In contradistinction to the ictal disorders of verbal behaviour described in the previous section, hypergraphia manifests itself between seizures.

The copious writing of hypergraphic patients is often replete with trivial notations and minute details. For instance, a patient of Waxman and Geschwind (1974) used to specify the time of the day at which the small events he recounted took place, *e.g.* "I took a walk this morning (8:30 to 9:30) when the sun was week (sic). This afternoon I had to take a nap for an hour (1:10 to 2:00)", or "I got home at 11:20, had lunch at 12:00 and worked on stamps in my bedroom all afternoon. I met X... at 5:15 and we came home."

Hypergraphic patients may write letters, diaries,

stories, poems, prayers, aphorisms, or lists. They may list such items as their friends, books, records, likes and dislikes, or the places they have been to. A patient of Waxman and Geschwind (1974) made up a list of the days between March 17 and 31 and May 1 and 16, 1970, each time indicating whether he had had a seizure or not. Absence of fits was recorded as "Thank 'GOD' none" or "I thank 'GOD' no seizures". For each attack, the time of its occurrence (*e.g.* 6.05 a.m.) and sometimes also its nature (*e.g.* "sexual fear seizure") was mentioned. When totalling the seizures for a given period, he expressed the numbers both in letters and in figures.

Hypergraphic patients may spend several hours a day writing and a number of them always keep a pencil and paper with them, so that they can write if they feel the need to. Some patients report that writing is compulsive and that they will write on whatever material is available. A patient of Roberts *et al.* (1982) used to produce scores of unconnected jottings. She would write on any piece of paper she could find. One of our own patients while in hospital wrote on the reverse side of old electroencephalograghic recordings. He used heaps of them during his stay in the clinic. Part of this extensive writing may be repetitive, some patients producing several times the same, or a similar, text. The writing may be supplemented with drawings, of which repeated copies are sometimes made.

Hypergraphic patients are not necessarily cultured individuals. Some of them do not fully master orthography and their notations are replete with spelling errors. As a matter of fact, a study by Jancar and Kettle (1984) shows that hypergraphia may be encountered in mentally handicapped epileptics. These patients may produce writing consisting of a few signs or letters repeated over and over again.

If the patient suffers from temporal lobe epilepsy he not infrequently deals with religious topics or religious experiences in his copious writing, and he may do so in his mother tongue as well in the foreign languages he is familiar with, as a case reported by Roberts *et al.* shows (1982).

Deblocking Epilepsy

While epilepsy generally renders verbal behaviour pathological, it may also, it would seem, improve verbal performance. Van Bogaert (1934) has reported the strange case of a 31-year-old patient who had post-encephalitic Parkinsonism. In addition, he had oculogyric crises. For the last three years, the patient had had no spontaneous speech. He communicated with his mother by means of gestures and eye movements. At times he appeared confused and evidenced severe comprehension difficulties. Curiously enough, the permanent mutism and the intermittent comprehension disorder both cleared during oculogyric crises. During these attacks the patient could use speech both receptively and expressively. However, he sometimes made palilalias. In this case, then, epilepsy appeared to play a deblocking role, temporarily relieving the patient of his speech impediment and comprehension difficulties. Van Bogaert did not discuss the possibility that mutism might have been psychogenic. In his view the abnormal brain activity that caused upward gaze deviation at the same time alleviated the organically based inhibition of the language centres.

Conclusions

This brief and incomplete review shows that paroxysmal brain activity may influence verbal behaviour in various ways. Conversely, verbal activity may precipitate abnormal bio-electric discharges in the brain. The study of epilepsy then confirms the view that there exist between language and the brain complex and in many respects surprising relationships, which challenge the sagacity of neuroscientists.

References

1. Alajouanine T, Nehlil J, Gabersek V (1959) A propos d'un cas d'épilepsie déclenchée par la lecture. Revue Neurologique 101: 463–467
2. Alajouanine T, Sabouraud O (1960) Les perturbations paroxystiques du langage dans l'épilepsie. L'Encéphale 49: 95–133
3. Benedek L (1925) Zwangsmäßiges Schreien in Anfällen als postencephalitische Hyperkinesie. Zeitschrift für die gesamte Neurologie und Psychiatrie 98: 17
4. Falconer M (1967) Brain mechanisms suggested by neurophysiological studies. In: Darley F, Millikan C (eds) Brain mechanisms underlying speech and language. Grune and Stratton, New York, pp 185–203
5. Foster F, Hansotia P, Cleeland C, Ludwig A (1969) A case of voice-induced epilepsy treated by conditioning. Neurology 19: 325–331
6. Herskowitz J, Rosman P, Geschwind N (1984) Seizures induced by singing and recitation. A unique form of reflex epilepsy in childhood. Archives of Neurology 41: 1102–1103
7. Hughlings Jackson J (1880) On a particular variety of epilepsy ("Intellectual aura"). One case with symptoms of organic brain disease. Brain 11: 179–207
8. Hughlings Jackson J, Colman WS (1898) Case of epilepsy with tasting movements and "dreamy state"—very small patch of softening in the left uncinate gyrus. Brain 21: 580–590
9. Jancar J, Kettle L (1984) Hypergraphia and mental handicap. J Ment Defic Res 28: 151–158

10. Lecours R, Joanette Y (1980) Linguistic and other psychological aspects of paroxysmal aphasia. Brain and Language 10: 1–23
11. Poeck K (1972) Stimmung und Krankheitseinsicht bei Aphasien. Archiv für Psychiatrie und Nervenkrankheiten 216: 246–254
12. Roberts J, Robertson M, Trimble M (1982) The lateralizing, significance of hypergraphia in temporal lobe epilepsy. J Neurol Neurosurg Psychiatry 45: 131–138
13. Souques A (1928) Note sur les troubles de l'écriture pendant les absences épileptiques et sur l'intérêt psychologique et médico-légal de ces troubles. Rev Neurol 353–360
14. Van Bogaert L (1934) Ocular paroxysms and palilalia. J Nerv Ment Dis 80: 48–61
15. Waxman S, Geschwind N (1974) Hypergraphia in temporal lobe epilepsy. Neurology 24: 629–636
16. Wohlfart G, Ingvar D, Hellberg A (1961) Compulsory shouting (Benedek's "klazomania") associated with oculogyric spasms in chronic epidemic encephalitis. Acta Psychiatr Neurol Scand 36: 369–377

Correspondence: Prof. Y. Lebrun, Ph.D., Neurolinguistics Department, School of Medicine, V.U.B., Laarbeeklaan, 103, B-1090 Brussels, Belgium.

VI. Aphasia and Personality

Acta Neurochirurgica, Suppl. 44, 113–117 (1988)

Aphasia and Personality

X. Seron[1] and **M. van der Linden**[2]

[1] Université Catholique de Louvain, Faculté de Médecine, Unité de Neuropsychologie expérimentale de l'Adulte NEXA, Bruxelles, Belgium,
[2] Hôpital de Bavière, Service de Neurochirurgie, Liège, Belgium

Summary

A critical review of the data in the literature and their interpretations is given. There are three main lines of interpretation:

— The neuropsychological point of view interprets personality disorders as a reaction to the language deficit;

— the anatomical point of view tries to relate personality disorders to the location and extent of brain lesions;

— the historical point of view links personality disorders to the pre-morbid personality.

All of them seem to have shortcomings and to be based at least in part on inadequate scientific methods including the principle problems of psychologically testing aphasics, of not having pre-morbid personality tests or using tests which are developed outside the field of neuropsychology.

We suggest the development of multifactorial approaches which take into account biological, sociological and psycho-individual variables in an interactive way.

Keywords: Aphasia; personality; affictive disorders.

The topic of aphasia and personality is obviously complicated by the vagueness and the extension of these two terms.

"Aphasia" is the more precise of the two at least on the diagnostic level. However there is more than one kind of aphasia, so we have to allow for the possibility that personality changes may vary with the nature and severity of the language and communication disorders.

As far as "personality" is concerned it has been noted that there are more than a hundred different definitions of what the term "personality" refers to. In view of the detailed discussion of Huber (this book) we will not go into this problem here. Instead we will adopt a minimalist position and use "personality" to refer to a structural and more or less permanent psychic organization that can only be inferred from the observation of regularities in the subject's reactions in different situations. This point of view clearly differs from a strict situationistic position since it implies that behaviour cannot be explained solely on the basis of "here and now" environmental variables. Structural variables have to be taken into account, even if it seems difficult at present to identify their nature, organization and process of realization (Huber 1977).

Yet, even if psychologists have not made significant progress in the comprehension of a concept of personality, they have clearly progressed in the causal analysis of affective and social adaptative problems.

In the past adaptative problems were considered the results of biological factors such as genetic predispositions or cerebral dysfunctions, of inadequate affective relations during infancy, or inadequate active social and environmental influences. Such mono-causal interpretations of behavioural disorders have been abandoned in favour of a multi-causal analysis of the responsible factors. At present, in clinical psychology the purpose is not to select one group of these factors as the critical one, but to understand how the psychological, environmental, and biological variables interact to provoke the appearance and the maintenance of structural changes in behaviour (Huber 1987).

Given this general evolution of clinical psychology, our main objective here is to review critically the little that has been published on personality changes in aphasic conditions to see if a similar evolution has occurred. Our evaluation will be largely negative, so we will suggest that no clear progress can be expected without a change of perspective and method.

As regards what has been done on personality disorders in aphasia, our first point is the rarity of the studies made and the second concerns their poor scientific quality.

These studies can be grouped along three main lines: one interprets personality disorders as a reaction to the language deficit (the neuropsychopathological point of view); another looks at personality disorders in relation to lesion characteristics (the localizationist point of view); and finally one that links personality disorders to the pre-morbid personality (the historical point of view).

Those who have suggested that personality changes result from the subject's reaction to his language deficit have contrasted anterior to posterior aphasics. Benson (1973, 1979) for example, considers that anterior aphasics are aware at least to some degree of their output speech deficit, that is, they know what to say but cannot say it. Consequently, these patients are both depressed and frustrated. When asked to perform difficult verbal tasks, they may have a serious emotional breakdown, which Goldstein called "the catastrophic reaction". While such catastrophic reactions are generally short-lived lasting only a few hours, Benson notes that they may continue for a number of days the patient often refusing to talk or to listen to anyone including members of their own family. Such depressive reactions may result in suicide, albeit rarely.

Language and affective problems in posterior and fluent aphasics, especially those with Wernicke's aphasia are very different. The crucial point is that such patients are frequently unaware of their own verbal deficits and often display abnormal unconcern. These patients are unable to monitor their verbal output and are also incapable of realizing that they produce jargon. Having a severe comprehension deficit, they are sometimes convinced that if they cannot understand the verbal messages addressed to them the fault is located in the speaker. The tendency to locate the blame outside the self may lead to an organized paranoid reaction. The patients consider that people are conspiring to speak in a mysterious code and not understanding what they say. According to Benson, such paranoid reaction may become dangerous and lead to physical attacks on hospital personnel, family members or other individuals.

Even if Benson considers that such differential reactions are caused by the site of the lesion, that is, the depressive reaction occurs with a pre-rolandic lesion and the paranoid maniac reaction with a post-rolandic lesion, the logic of his interpretation is clearly psychological. Personality changes are interpreted as an adaptative response of the subject to specific and different cognitive and communication problems. Wernicke's aphasics could develop paranoid reaction because of their comprehension deficit and lack of awareness, Broca's aphasics depression because of their speech output problems and their awareness of the deficit. Other neurolinguists have advanced more or less similar interpretations. For example, Ducarne (1986) also suggests that different personality changes occur in relation to the different types of aphasia.

These clinical descriptions have received some support from the more systematic study of Gainotti (1969, 1972) of a group of 150 patients with unilateral lesions, in whom a high incidence of catastrophic reactions was observed in those patients with serious communication problems, especially when they were confronted with difficult verbal tasks. Nevertheless, such psychological interpretations have several shortcomings.

The first one is *the specificity* of the descriptions furnished. Are these psychological reactions typical to aphasics or could they appear in other pathological conditions?

The question of specificity can only be resolved by carefully examining what occurs in selected control groups, but then how could one define or find an adequate control group?

One solution would be to compare aphasics with other patients with brain lesions. This has, for example, been done by Prigatano (1987) who observed catastrophic reactions in patients with and without aphasia, and by Robinson *et al.* (1984) who observed more frequent depressive reactions after left pre-Rolandic lesions but independently of aphasic disorders.

Another possibility consists in comparing aphasics with subjects with no brain lesions with an acquired verbal deficit. Such research has not been done in neuropsychology, and, to our knowledge no studies have been done in neuropsychology comparing psychological reactions to different acquired verbal handicaps such as acquired deafness, laryngectomy, tongue injury, and aphasia. Such comparisons are critical for establishing the specificity, if any, of the aphasic reactions and would be interesting in themselves for clinical psychologists. Just as an example, it has been observed that acquired deafness or experimentally induced hearing deficits may lead to paranoid reactions, which could be similar to the reactions of some Wernicke's aphasics (Zimbardo *et al.* 1981).

The second difficulty, which is not specific to this research orientation is the problem of *inter-subject variability*. Clinical descriptions as well as the studies of Gainotti and Robinson show that not all Wernicke's or all Broca's aphasics present personality disorders. Furthermore, if the intensity of the affective reaction depends on the degree of severity of the communication disorders, then personality changes would be proportional to the language handicap. This point has never been systematically studied and, at least on intuitive grounds, the existence of such a linear relation does not seem very plausible.

The third difficulty concerns *the methodology*. These clinical descriptions were presented without sufficiently clear indications about the procedure used.

Our main criticisms concern:

— the criteria used for labelling a reaction "catastrophic" depressive, paranoid, and the like were not specified,

— the methodology used: if the reaction has been directly observed, was it by at least two independent observers? Were there criteria of agreement between the judges?

— In what context: at home, during testing, or in therapy?

If the patient's reactions were not directly observed, were they reported by the patient or by relatives?

The data collected by McKinlay and Brooks (1984) on head-injured patients support the view that there are systematic differences between the patient's and the relative accounts of emotional and behavioural changes and that the relative's personality influences their perception of changes in the patient. For example, the neuroticism scores of the relatives correlated significantly with the emotional aspects of their reports. The age of the relatives may also have an effect: elderly spouses have been shown to be more sensitive to the physical problem of the aphasic patient, while young spouses seems to be more sensitive to relational and communication deficits (Artes and Hoops 1976). In fact, methodological aspects such as the conditions of observation and the personality, age, and status of the observer have proved to be crucial in data collection in clinical psychology. For a more detailed discussion we refer to McKinlay and Brooks (1984).

In summary, the existing clinical descriptions of the reaction of aphasics to their language disorders cannot be viewed as scientific facts, because the methodology is inadequate, the specificity of reaction has not been established and the causes of intersubject variability has not been determined.

The second and actually more prevalent orientation of research holds that personality changes in aphasics have to be considered the result of the inter- and intra-hemispheric location of the lesion.

At the origin of this research orientation is the work of Gainotti (1969, 1972), who observed in a group study a prevalence of catastrophic reactions after left hemisphere lesions and cheerful acceptance of the disability, a tendency to joke, and indifference to or denial of illness after right hemisphere lesions. Gainotti considered the catastrophic reactions to be dramatic but psychologically appropriate while he considered the "indifference reaction" to be an inappropriate form of emotional behaviour that he attributed to the disruption of a structure critically involved in emotions and affect presumably located in the right hemisphere.

This original work has been the impetus for numerous studies on the so-called hemispheric laterality of emotions. The main areas of research are:

1) analysis of emotional behaviour with unilateral brain lesions (Heilman *et al.* 1984);

2) experimental research conducted both in brain-damaged patients and in normal subjects to study the processing of emotional information and the expression of emotion (Campbell 1982);

3) investigations conducted in patients with right and left temporal lobe epilepsy (McIntyre *et al.* 1976, Bear and Fedio 1977);

4) studies of psychiatric patients (Gruzelier and Flor-Henry 1979, Coffey 1987).

Two main interpretations of the relationships between emotions and cerebral dominance have resulted and are still under discussion in the literature:

— The first one assumes an overall dominance of the right hemisphere for various kinds of emotions and affects (Gainotti 1969, 1972),

— the second, on the contrary, proposes a right hemisphere dominance for negative affect and a left hemisphere dominance for positive affect (Sackeim *et al.* 1978, Sackeim and Gur 1978, Schwartz *et al.* 1979, Bruyer 1980) *.

Given the restricted concern here, we will only examine some of the implications of these studies for personality changes in aphasics.

In the perspective that assumes that affective changes are due to the location of the lesion, the critical point is to demonstrate the presence of personality change in the absence of aphasia or that personality changes are not correlated with some characteristics of the language and cognitive disorders.

This is the perspective adopted by Robinson and others (1984) who have, by means of successive studies and standardized personality tests and questionnaires,

* A third position is to consider that the differential contribution of the cerebral hemispheres in emotional behaviour remains an open question. The problems not yet resolved are the interpretation of the observed lateral differences that could relate either to the emotional nature of the process or to the involved perceptual or motor components (Feyereisen 1987).

examined personality changes in relation to localization of brain lesions evaluated by means of CT-scan.

The main results of these authors may be summarized as follows:

— depression scores are significantly higher with left lesions than with right lesions;

— depression scores are higher with anterior left lesions than with any other left or right location;

— in the group of patients with left anterior lesions, depression scores correlate with the proximity of the lesion to the frontal lobe;

— in the group with left anterior lesions, lesion volume correlates with the severity of depression;

— depression in patients with left anterior lesions is independent of the presence of language disorders.

These authors thus found a graded relationship between depression scores and the proximity of the lesion to the left frontal lobe, which in their opinion, cannot be explained by the patient's reaction to cognitive or language deficits. They then speculate on the role in depression of a disruption of noradrenergic fibres that could be more important in the anterior cortex.

Whatever the interest of such observations, the work of Robinson and his colleagues is not without limitations that raise interesting methodological problems. Since they used various standardized tests that require patient collaboration, 53 out of their 184 patients, especially those with a severe aphasia comprehension deficit, could not be interviewed. The anterior/posterior left gradient must, therefore have been biased as well as the influence of aphasia on depression. Furthermore divergent data have appeared in the literature; Grafman and collaborators in 1986 have shown increased depression scores in a group of Vietnam veterans with right orbital lesions as compared with left ones. The hemispheric lateralization of the cerebral mechanisms that accompany depression thus requires more documentation.

Whatever the issue of this dispute the Robinson study has clearly shown that depression may occur in brain damaged patients without aphasic disorders. Thus even if the site of the lesion is not the whole story there is sufficient neuropsychological evidence that it plays a critical role in dramatic emotional reactions (Poeck 1969), and as the reception and expression of emotions are part of the concept of personality, lesion characteristics may well constitute a critical variable.

The third orientation is twofold:

On the one hand one asks if the pre-injury personality influences the post-injury personality changes. This direction of research was developed by Kozol (1945, 1946) whith traumatic patients, but to our knowledge no study with aphasics has systematically controlled this variable.

On the other hand, it has been suggested that the premorbid personality influences the expression of language and cognitive disorders. Weinstein and Lyerly (1976) have interpreted the classic Wernicke's jargon with motor agitation as a kind of defense mechanism elaborated by patients presenting specific premorbid personality traits, but such an interpretation has never been the object of systematic control. We note, however, that the idea that affective reactions may be an intrinsic part of the language disorders has recently resurfaced in the neurolinguistic literature and constitutes a critical elements of some reinterpretation of the agrammatism semiology by Heeschen (1985) and Kolk (1985).

The main difficulty with such an approach is the evaluation of the premorbid personality. In general patients have never had any personality tests before their brain injury. The observer is thus obliged to infer through interviews and testimony what the premorbid psychic organization was. This is, of course, very difficult, and the risk of errors is high and not easily controllable.

In summary none of these three main research orientations have been able to produce solid data. Some of the difficulties are methodological. Studies are difficult to compare given the absence or the variability of the patient selection criteria. The instruments used to evaluate emotional and personality changes also raises many problems. In clinical observations it is difficult to know precisely what labels such as irritability, anxiety, euphoria, depression and so on refer to and the context in which the behaviour was observed.

The use of personality tests such as the MMPI (Minnesota Multiphasic Personality Inventory), the Rorschach test, and the Thematic Apperception Test (TAT), which were developed outside the field of neuropsychology is also problematical.

— First, they are not practicable with all patients, and when they are, it is difficult to discriminate in the subjects responses what is due to cognitive and language deficits from what indicates personality changes.

— Second, the use of nosographic classification from psychiatry may on *a priori* grounds bias the interpretation. In recent years, more rigorous and systematic studies using behavioural scales have appeared (Bear and Fedio 1977, McFinlay and Brooks 1984, Grafman *et al.* 1987, Robinson *et al.* 1984). However the theoretical foundations of these scales remain vague and this points to the inescapable poverty of the present psychological model of personality.

— Finally, the role of sociological variables, pre-, per-, and post-injury has been underestimated. These variables are manifold: socio-economic level, professional achievement, familial role and position, institutional, medical and psychotherapeutic setting. All of these variables have proved to influence the patient's cognitive recuperation but their impact on the aphasic personality has not been closely examined (Seron 1980).

Thus it seems necessary to develop multifactorial approaches that take into account biological, sociological, and psycho-individual variables in an interactive way. It seems also necessary on a theoretical level to develop more precise hypotheses on the relationship between cognitive and affective disorders, as has been done by Johnson and Magaro (1987) in their attempt to relate affective and memory problems in depression.

Such multifactorial and interactive approaches may be conducted either on group studies or in single-case monography as has recently been done by Leftoff (1983) who set out to describe the development of a paranoid reaction in one aphasic patient by postulating precise relationships between cognitive deficits and paranoid reactions. Furthermore as no source of information may be discarded one must also examine the retrospective testimony of cases of reversible and paroxysmal aphasia.

References

1. Artes R, Hoops R (1976) Problems of aphasic and non-aphasic stroke patients as identified and evaluated by patient's wives. In: Lebrun Y, Hoops R (eds) Recovery in aphasics. Swets & Zeitlinger, Amsterdam
2. Bear DM, Fedio P (1977) Quantitative analysis of interictal behaviour in temporal lobe epilepsy. Arch Neurol 34: 454–467
3. Benson DF (1973) Psychiatric aspects of aphasia. Br J Psychiatry 123: 55–566
4. Benson DF (1979) Aphasia, alexia and agraphia. Churchill Livingstone, New York
5. Bruyer R (1980) Implication différentielle des hémisphères cérébraux dans les conduites émotionelles. Acta Psychiatrica Belgica 80: 266–284
6. Campbell R (1982) The lateralization of emotion: a critical review. Inter J Psychology 17: 211–229
7. Coffey EC (1987) Cerebral laterality and emotion: the neurology of depression. Comprehensive Psychiatry 28: 197–219
8. Ducarne B (1986) Rééducation sémiologique de l'aphasie. Masson, Paris
9. Feyereisen P (1987) Non verbal Communication. In: Rose FC (ed) Aphasia. Whurr, London
10. Gainotti G (1969) Réactions "catastrophiques" et manifestations d'indifférence au cours des atteintes cérébrales. Neuropsychologia 7: 195–204
11. Gainotti G (1972) Emotional behaviour and hemispheric side of lesion. Cortex 8: 41–55
12. Grafman J, Vance SC, Weingartner H, Salazar AM, Amin D (1986) The effects of lateralized frontal lesions on mood regulation. Brain 109: 1127–1148
13. Gruzelier J, Flor-Henry P (1979) Hemisphere asymetries of function in psychopathology. Elsevier, Amsterdam
14. Heeschen C (1985) Agrammatism versus paragrammatism: a fictitious opposition. In: Kean ML (ed) Agrammatism. Academic Press, New York
15. Heilman KM, Watson RT, Bowers D (1984) Affective disorders associated with hemispheric disease. In: Heilman KM, Satz P (eds) Neuropsychology of human emotion. Guilford, New York
16. Huber W (1977) Introduction à la Psychologie de la personnalité. Mardaga, Bruxelles
17. Huber W (1987) La psychologie clinique aujourd'hui. Mardaga, Bruxelles
18. Johnson MH, Magaro PA (1987) Effects of mood and severity on memory processes in depression and mania. Psychological Bulletin 101: 28–40
19. Kolk HHJ, van Grunsven A (1985) Agrammatism as a variable phenomenon. Cognitive Neuropsychology 2: 347–384
20. Kozol H (1945) Pretraumatic personality and psychiatric sequelae of head injury. Arch Neurol Psychiat 358–364
21. Kozol H (1946) Pretraumatic personality and psychiatric sequelae of head injury. Arch Neurol and Psychiat 56: 245–275
22. Leftoff S (1983) Psychopathology in the light of brain injury: a case study. J Clin Neuropsychology 5: 51–63
23. McIntyre M, Pritchard PB, Lombroso CT (1976) Left and right temporal lobe epileptics: a controlled investigation of some psychological differences. Epilepsia 17: 377–386
24. McKinlay WW, Brooks DN (1984) Methodological problems in assessing psychosocial recovery following severe head injury. J Clin Neuropsychology 6: 87–99
25. Poeck K (1969) Pathophysiology of emotional disorders associated with brain damage. In: Vinken PJ, Bruyn AW (eds) Handbook of clinical neurology. North Holland, Amsterdam
26. Prigatano GP (1987) Personality and psychosocial consequences after brain injury. In: Meier MJ, Benton AL, Diller L (eds) Neuropsychological rehabilitation. Churchill Livingstone, New York
27. Robinson RG, Kubos KL, Starr LB, Rao K, Price TR (1984) Mood disorders in stroke patients. Importance of location of lesion. Brain 107: 81–93
28. Sackeim HA, Gur RC (1978) Lateral asymmetry in intensity of emotional expression. Neuropsychologia 16: 473–481
29. Sackeim A, Greenberg MS, Weiman L, Gur RC, Hungerbuhler JP, Geschwind N (1982) Hemispheric asymmetry in the expression of positive and negative emotions. Arch Neurol 39: 210–218
30. Schwartz GE, Ahern GL, Brown SL (1979) Lateralized facial muscle response to positive and negative emotional stimuli. Psychophysiology 16: 561–571
31. Seron X (1980) Aphasie et neuropsychologie: approaches thérapeutiques. Mardaga, Bruxelles
32. Weinstein EA, Lyerly OG (1976) Personality factors in jargon aphasia. Cortex 12: 122–133
33. Zimbardo P, Andersen S, Kabat L (1981) Induced hearing deficit generates experimental paranoia. Science 212: 1529–1531

Correspondence: Prof. X. Seron, Ph.D., Université Catholique de Louvain, Faculté de Médecine, Unité de Neuropsychologie, expérimentate de l'Adulte, NEXA, Avenue Hippocrate, 5545, B-1200 Bruxelles, Belgium.

VII. Psychosurgery and Personality

Acta Neurochirurgica, Suppl. 44, 121–124 (1988)

Psychosurgery and Personality Disorders

P. Cosyns

University of Antwerp (UIA), Department of Psychiatry, Wilrijk, Belgium

Summary

Defining personality and its pathological variants is a hazardous enterprise. The personality concept refers to the global coherence of functioning of a person as a whole and can be divided into two components, temperament and character. Personality disorders will be discussed according to the DSM III-R (1987) (The Diagnostic and Statistical Manual of Mental Disorders) the most recent psychiatric taxonomy of the American Psychiatric Assocation. The interaction between psychosurgery and personality is multiple. It will be stated that the mere presence of a personality disorder in a patient should never be an indication for psychosurgery. It may sometimes even act as a contraindication. Psychosurgery can produce changes in some basic psychic dysfunctions and although there is no universally accepted understanding of how it works, some hypothetical neurobiological foundations will be discussed.

Keywords: Psychosurgery; personality disorder; DSM III-R.

1. The Concept of Personality

1.1. The statement that human beings are constitutionally different is no longer controversial. Child psychology provided ample supporting empirical evidence that individual children differ markedly from one another in a large number of different ways. These differences can be measured with reasonable reliability. It has also be well demonstrated that these differences influence the further development, shape the personality style, and are relevant for later development of psychiatric disorders (Rutter 1987). But the literature displays a surprising lack of conceptual uniformity about terms such as "personality", "temperament", "character" and "personality disorder". Clarification of these terms is necessary since they refer to different phenomena and processes.

The personality concept refers to the global coherence of functioning of a person as a whole. The emphasis lies on integrative and organizing functions of cognitive, affective, motivational as well as overt behavioural components (traits) that persist over an extended period. Personality concerns deeply embedded meaningful patterns of behaviour and involves a crucial historical dimension: it can somehow be stated that the history of the personality is the personality.

1.2. The basic components of the personality structure are several individual prefered patterns of behaving which become very resistant to extinction. Personality "traits" is generally used in the literature to designate those more simple characteristic basic patterns of behaving. The DSM III-R (1987) refers to personality traits as enduring patterns of perceiving, relating to, and thinking about the environment and oneself who are exhibited in a wide range of important social and personal contexts. These personality traits characterize the persons recent and long-term functioning and are by no means pathological by themselves.

1.3. The term "pattern" in defining personality means that the basic traits, dispositional attitudes or characteristics are not criss-crossed at random but organized in a predictable structure of overt and covert behaviours. In this perspective, three of the actual main approaches to personality and its assessment will be briefly mentioned.

The progressive shaping of social and interpersonal relationships through the biographical history of the person is the first important approach. The intensive analysis of the individual biographical development allows us to describe a shaping process of a set of behaviours mainly designed to achieve reinforcement and avoid punishment. Those early learned and often practised specific behaviour responses are in some sense internalized and act as an "internal working model". They shape future relationships or attachments and so they have a predictive value. On the basis of their earlier relationships children shape and differentiate their later relationships. The early parent-child attachment acts as a model for the development of a set of expectations about their own relationship capacities and other peoples response to it.

This progressive shaping of social and interpersonal relationships implies some development of cognitive concepts about relationship and links closely with the second main approach to personality, namely the cognitive self-concepts. Personality involves a set of cognitions about ourselves, relationships and interactions with the environment. The main idea is that we function according on our concept about what sort of people we are. In the same way we respond to others not only on the basis of their actual behaviour, but on the basis of our perception of what sort of a person they are. Cognitive self-concepts are flexible and dynamic, and consequently by no means static as the traditional personality traits are generally conceptualized.

The strategy of direct behavioural assessment is the third actual approach briefly mentioned in this review. Personality theories try to rely more heavily on still sparse but rapidly growing empirical research data. There is a greater focus on what the person does in specific situations without assuming *a priori* that the actual overt behaviour is the distorted manifestation of an underlying inferred cause or structure. Behavioural asessment is most useful in evaluating the specific effects of any given treatment.

1.4. Personality can be divided into two main components or subsets: temperament and character. Temperament refers to a rather small number of simple features of the disposition which may be viewed as being the basic biogenic foundation of personality. Newborns display distinct responses which differentiate them from birth on, before any environmental learning experience could account for them. For example, some newborns suck energetically while others are more apathetic.

These basic temperamental traits appear early in childhood, show a great stability over time, have a direct neurobiological correlate and constitute a predictable distinctive pattern of responses. They are "simple" and "basic", which means "non-cognitive" and "non-motivational" such as for example activity, emotionality, sociability and impulsivity (Rutter 1987).

Character reflects more nurture and the persons learning from upbringing, society and culture. The personal adherence to the ethics, social mores and cultural societal values are the main constituent elements of the character. It reflects necessarily a moral judgement on a person's behaviour.

2. Personality Disorders and Their Classification

2.1. The differentiation between normal and abnormal personality remains still a harzardous enterprise. The so-called personality disorders have a very tumultuous history in psychiatry but four unifying features can be recognized:

— an early onset in childhood or adolescence and continuation throughout most of adult life,

— the long-standing persistence over time without marked remission or relaspes, although some settling of the most disturbing symptomatology may be seen in middle or old age,

— early onset and persistence of abnormalities that constitute a basic aspect of the persons usual functioning,

— abnormalities that cause either significant impairment in social and occupational functioning or subjective distress.

Normal and abnormal personalities can best be conceived as being separate points along a same and unique continuum. The difference between them is a quantitative one of degree and not a qualitative one of a difference in state or condition. Millon (1981, 1985) elaborated some useful clinical criteria to distinguish the abnormal personality pattern from their normal variants: adaptative inflexibility, a tendency to foster vicious circles and tenuous stability.

Adaptative inflexibility refers to the limited ways of coping with the environment. A restricted behavioural repertoire is used rigidly and inappropriately upon situations for which they are ill-suited. Furthermore pathological personality patterns are themselves pathogenic by intensifying pre-existing difficulties. Thirdly, abnormal personality patterns are fragile and "decompensate" easily under conditions of subjective stress or environmental pressure (weak ego-strength).

2.2. The classification of personality disorders is another controversial issue in contemporary psychiatry. The dispute as to the therapeutic usefulness of categorization and diagnostic labels will not be reviewed here, considering the evident necessity of a reliable classification system for research purposes.

The Diagnostic and Statistical Manual of Mental Disorders (DSM III-R 1987) of the American Psychiatric Association fulfills this requirement.

The main noteworthy features of this classification system are the following:

— a resolute choice for a descriptive approach and description of the disorders without aetiological inference,

Table 1

	Personality disorders		
Clusters DSM III-R	odd eccentric	erratic dramatic emotional	anxious fearful
Pers. dis. DSM III-R	paranoid schizoid schizotypal	histrionic narcissistic antisocial borderline	avoidant dependent compulsive passive-agressive
Eysenck	psychoticism	extroversion	neuroticism
Psychophysiol. Correlate (Siever et al. 1983)	vulnerability to cognitive disorganisation	low cortical arousal and disinhibited autonomic or motor activity	high cortical arousal with motor inhibition

— the elaboration of specific diagnostic criteria as guides for making each diagnosis since such criteria obviously enhance interjudge diagnostic reliability. In consequence, each personality disorder is defined by a set of identifiable criteria which requires as little inference as possible on the part of the observer. Examination of the criteria for the various categories of personality disorders reveals a rather disturbing level of overlap, which strengthens the view of personality disorders as being dimensional rather than categorical.

— a multiaxial evaluation system along five axes provides a global biopsychological approach to assessment: Axis I comprise all the mental disorders, Axis II personality disorders, Axis III physical disorders and conditions, Axis IV severity of psychosocial stressors and Axis V global assessment of functioning.

Axis II personality disorders have been grouped into three clusters (Table 1). People with the first cluster appear odd, eccentric and fearful of all social relationships. The second cluster reflects a tendency to be extroverted, emotional and social with a low acting-out threshold. Persons with the third cluster are introverted, anxious, fearful and provide a tendency to turn anger against the self. As will be seen those different clusters are relevant as indications for psychosurgery.

The three basic personality dimensions of Eysenck's theory can be linked with the DSM III-R clusters: psychoticism (cluster I), extroversion (cluster II) and neuroticism (cluster II).

3. Personality Disorder and Psychosurgery

3.1. Several aspects of the actual interference between personality disorder and psychosurgery will be discussed. Is the mere presence of a personality disorder an indication for surgery? On the contrary, can the presence of a personality disorder constitute a contraindication? Does psychosurgery change the personality of the patient? Answers to these crucial questions will partially rest on data from the literature but mainly on the authors experience as a member of a "committee on psychosurgery". Representatives of relevant specialisms—*i.e.* psychiatry, neurosurgery and clinical neurophysiology—evaluate in this committee indications and contraindications for psychosurgery. During a ten year period, 71 psychiatric cases were submitted to this committee by individual psychiatrists or psychiatric treatment teams in the Netherlands and Belgium (Storm van Leeuwen 1982, Cosyns *et al.* 1988).

3.2. It is obvious, according to the actual state of the art, that the mere presence of a personality disorder never should constitute an indication for psychosurgery. Changing the patients personality is not the primary aim of a psychosurgical intervention. Psychiatrists, patients and their relatives are rather afraid of possible personality changes that could be considered as unwanted side-effects. Only some well defined Axis I disorders of the DSM III-R and never Axis II disorders can be taken consideration for psychosurgery. Those disorders—for example, obsessive compulsive disorder, major depression, impulse control disorder—are complex clinical constructs with an even complex multifactorial, not fully understood aetiology and pathogenesis. To think about psychosurgery as a simple mean for curing such conceptually complex disorders is giving proof of naivety. The targets of psychosurgery are most frequently therapy resistent and disturbing anxiety linked behaviours, such as obsessions or compulsions, and depressive mood or aggressive behaviour. Some basic behavioural features about anxiety, depressive mood or aggression can definitely be surgically modified.

3.3. Patients taking psychosurgery into consideration as a possible form of treatment for their Axis I disorder, can—and frequently do—present concomitantly an Axis II diagnosis, *i.e.* a personality disorder. Patient's pathological personality structure will never as such be, a target for psychosurgery, but it may influence the response to the operation. Considering the three already mentioned personality clusters, our experience with frontal lobe targets can be summarized as follows: the anxious/fearful pattern responds very well to psychosurgery, the erratic/dramatic/emotional rather badly and the third cluster, the odd/excentric one, remains apparently uninfluenced. The existing personality structure can influence the decision—making process for psychosurgery. For the clinician it means for example, that the presence of an outspoken histrionic personality (second cluster) in an obsessive compulsive patient will constitute a contraindication for operation. The same patient with a personality disorder of the anxious/fearful cluster will on the contrary receive positive advice for surgery, provided that other criteria about his obsessional syndrome are also met. An existing personality disorder is never an indication for psychosurgery but can be contraindication.

3.4. There is no universally accepted understanding of how psychosurgery works, but obviously it sometimes produces change and reduces certain biological vulnerabilities of the patient. According to our own experience with 21 operated patients with obsessive compulsive disorder, the following observations were made (Cosyns *et al.* 1988). An immediate decrease of the anxiety level with an increased sensitivity to anxiolytics and antidepressants, improvements in social behaviour, increased interpersonal assertivity and an increased capacity to experience internal feelings and express emotions. Postoperative psychological tests failed to reveal any evidence of organicity. Incidental transient negative effects and complications were, postoperative confusion, disinhibited behaviour, transient ataxia and epileptic seizures occurring several years after the operation (mainly during an episode of alcohol abuse). Those changes do not involve the complex cognitive or motivational components of personality, but rather the more basic dispositional or temperamental traits such as mood, anxiety, activity, impulsivity ... Psychosurgery influences these basic features which hypothetically must have a fairly direct neurobiological correlate.

The biosocial personality theory of Cloninger (1987) offers some interesting suggestions.

He isolated three genetically independent personality dimensions linked to a different monoamine neuromodulator: novelty seeking (behavioural activation) and dopamine, harm avoidance (behavioural inhibition) and serotonin, reward dependance (behavioural maintenance) and noradrenaline. A more scientific approach to psychosurgery could surely shed more light on the neurobiological foundations of some basic psychic dysfunctions involved in more complex states such as personality disorders or clinical psychiatric syndromes.

Acknowledgements

I would like to acknowledge the help of W. P. Haaijman, J. A. Ceha, No Sijben, J. Gybels, J. van Manen, C. van Veelen, all members of the interdisciplinary "Committee on Psychosurgery" (Utrecht, the Netherlands), and Miss Rita Herremans for the typescript.

References

1. Cloninger RC (1987) A systematic method for clinical description and classification of personality variants. Archives General Psychiatry 44: 573–588
2. Cosyns P, Haaijman W, Ceha J, Sijben N, Gybels J, van Manen J, van Veelen C, Follow-up study of patients with obsessive compulsive disorder treated by psychosurgery. (Submitted for publications)
3. Diagnostic and statistical manual of mental disorders (1987), 3rd ed. American Psychiatric Association
4. Millon T (1981) Disorders of personality DSM III: Axis II. John Wiley and Sons
5. Millon T, Everly GS (1985) Personality and its disorders, a biosocial learning approach. John Wiley and Sons
6. Rutter M (1987) Temperament, personality and personality disorder. Br J Psychiat 150: 443–458
7. Siever LJ, Insel TR, Uhde TW (1983) Biogenic factors in personalities. In: Frosch JP (ed) Current perspectives on personality disorders. American Psychiatric Press, Washington D.C.
8. Storm van Leeuwen W (1982) Neuro-physico-surgery in the Netherlands since 1971. Acta Neurochir (Wien) 61: 249–256

Correspondence: P. Cosyns, M.D., University of Antwerp (UIA), Department of Psychiatry, Universiteitsplein 1, B-2610 Wilrijk, Belgium.

Acta Neurochirurgica, Suppl. 44, 125–128 (1988)

Historical Overview of Psychosurgery and Its Problematic

H. Th. Ballantine, Jr.

Harvard Medical School, Massachusetts General Hospital, Boston, U.S.A.

Summary

In 1936, Egas Moniz published his first paper on frontal leucotomy for psychiatric illness. The initial enthusiasm for this innovative treatment of intractable psychiatric disorders and chronic pain was soon tempered by reports of undesirable side-effects, and neurosurgeons began a search for modifications of leucotomy which would increase safety without reducing efficacy.

As a result of these clinical investigations, the original imprecise, radical frontal leucotomy has been superseded by precise, small stereotactically placed lesions in the limbic system and the descriptive phase "limbic system surgery" is replacing the out-moded word "psychosurgery".

This presentation will document some of the more important contributions to a currently under-utilized, often criticized approach to the treatment of suffering individuals chronically disabled by psychiatric illness and pain.

Keywords: Cingulectomy; cingulotomy; leucotomy; lobotomy; psychosurgery.

In the summer of 1935 the Second International Congress of Neurology was convened in London, England. In attendance were two men from entirely different cultural and professional backgrounds who were destined to inspire and inaugurate neurosurgical procedures for the treatment of psychiatric illness: John Fulton and Egas Moniz.

Fulton, professor of physiology at Yale University's School of Medicine, participated in a symposium on frontal lobe function and spoke on the changes in behaviour that he and his psychologist co-worker, Carlyle Jacobsen, had observed in two female chimpanzees that had been subjected to bilateral frontal lobectomies. Most remarkable was the finding that these animals could still carry out most of the complex tasks for which they had been trained but were significantly changed in certain responses to their environment. They had become less fearful of humans and more tractable. They appeared to have a more placid attitude toward situations (such as being unable to carry out tasks for which they were rewarded by food) that had preoperatively caused frustrational behaviour, reactions which Fulton and Jacobsen termed "experimental neurosis".

Egas Moniz, Portugal's most distinguished neurologist and come to the Congress to present an exhibit of his pioneering investigations of cerebral angiography using thorotrast. By a coincidence that was to have an important bearing on the development of psychosurgery his exhibit adjoined one prepared by Walter Freeman on the use of Thorotrast for ventriculography. Freeman held appointments in neurology at Georgetown and George Washington University Medical Schools in Washington D.C. as well as posts in neuropsychiatry and neuropathology at St. Elizabeth's, the federal government's mental hospital in nearby Anacostia, Maryland. He had made the acquaintance of Egas Moniz at the First International Congress of Neurology four years earlier; the proximity of their two exhibits caused this acquaintanceship to ripen into a friendship and they attended the symposium on frontal lobe function together.

At the conclusion of Fulton's lecture Egas Moniz told Freeman that he thought it might be possible to treat psychiatric illness by placing lesions in the frontal lobes. Freeman was intrigued by his bold suggestion and asked to be kept informed if he did indeed decide to investigate this idea.

Egas Moniz returned to Lisbon in August of 1935 and immediately set about assembling a team consisting of a young neurosurgeon, Almeida Lima, a neurologically oriented clinician, and two psychiatrists one of whom, Sobral Cid, was professor of psychiatry at the University in Lisbon. Three months later he and Lima carried out their first lobotomy by injecting alcohol in the white matter of the frontal lobes. This technique was used in three additional patients before the two investigators moved to the use of a "leucotome" to sever volumes of white matter about 1 cm in diameter; several such lesions were placed in each frontal lobe.

In June of 1936 Egas Moniz published his now classic paper describing the results of these surgical procedures in the first 20 patients: 7 were felt to be greatly benefited, another 7 were helped and the psychiatric status of the remaining 6 was unchanged. All of these patients were hospitalized in psychiatric institutions and diagnosed as having incurable psychiatric illnesses[3].

Walter Freeman was fascinated by the work of Moniz and he, in turn, sought the assistance of a neurosurgeon, James Watts. Together they carried out the first American frontal lobotomy on September 14, 1936. During the ensuing 5 years Freeman and Watts operated upon 108 patients with four postoperative deaths[4]. Seventy four cases were described in great detail and the authors felt that 63% of had sustained worth-while benefit; 23% were relatively unchanged and 14% (including the fatalities) were made worse. In this last group, the living patients were described as having a "frontal lobe syndrome" characterized by nonchalance, inability to carry out tasks and loss of social control.

These complications, recorded by others, led to an exploration of the possibility that more restricted lesions might reduce the incidence of undesireable side effects and yet be just as beneficial as radical pre-frontal lobotomy. As a result, the original operation of Egas Moniz was abandoned in favour of restricted frontal leucotomies, three of which (bilateral inferior leucotomy, bimedial leucotomy and orbital gyrus undercutting) were done most frequently throughout the world. Nevertheless, the complication rate, while reduced, was of sufficient concern to lead to another significant step toward modern psychiatric surgery.

At the 1947 meeting of the Society of British Neurological Surgeons there was a general discussion of the problems associated with the current operative approaches. At one point John Fulton, who was present as a guest of the Society, remarked that, were it feasible, the cingulum might be a safer region in which to place lesions with results similar to those made in the frontal lobes since, "In cases which have been deemed favourable and have come to autopsy, the anterior cingulate has been involved almost invariably ..."[5]. Sir Hugh Cairns of Oxford and Professor Jacques le Beau from Paris heard this statement and both proceeded promptly to act on it. In 1961 Lewin reviewed the status of 26 cingulectomized operated upon by Cairns; 21 were said to have benefited and 11 of them were classified as greatly improved[8].

In the late 1940s two other events occurred which, over the next decade, completely transformed the still relatively crude operations into the various modern, precise procedures that are performed today: the development of the limbic system concept and the introduction of stereotaxis.

James W. Papez published in 1937 his now classic paper "A Proposed Mechanism of Emotion" in which he theorized that the hypothalamus, the gyrus cinguli, the hippocampus and their interconnections, formed a neuroanatomic and neurophysiologic unit which was the fundamental "basis of the emotions"[11]. Not much attention was given to this theory until it was enthusiastically embraced by Paul Maclean ten years later. In 1952 Maclean coined the now familiar term "limbic system" to encompass the Papez construct as well as the other structures which had come to be identified with it.

By 1948 psychosurgery had become so well accepted that the First International Congress of Psychosurgery was convened in Lisbon to honor Egas Moniz who, a year later was awarded the Nobel Prize for his introduction of prefrontal lobotomy. During the congress, Spiegel and Wycis described their "stereoencephalotome" a stereotactic device modeled after the apparatus used by Horsley and Clark forty years before for animal research[12]. The inauguration of stereotactic surgery now permitted neurosurgeons to reach any structure within the limbic system for the treatment of psychiatric illness.

Although interest in stereotactic surgery grew rapidly during the 1950s, it was not applied with great frequency to the limbic system. Actually, interest in psychiatric surgery waned in this period as a result of the introduction of psychotropic drugs; an increasing number of psychiatrists believed that these combined with electroshock therapy would prove to be the cure for otherwise intractable disorders of affect and the amelioration of the manifestations of schizophrenia. It was gradually realized however, that neither of these modalities alone or in combination were the complete answer. Moreover, complications resulting from the use of psychoactive drugs, particularly the tardive dyskinesias, were seen with increasing frequency. These factors lead in the 1960s to a revival of interest in psychiatric surgery. It was during this decade that stereotactic approaches to the limbic system increased in popularity, and the use of psychiatric surgery was again in the ascendancy. It soon became apparent that another international meeting would be valuable.

The Second International Congress of Psychiatric Surgery was held in 1970 in Copenhagen, Denmark; 51 papers were contributed by 77 authors from 11 countries. One of the most important reports was given by Knight[7] regarding 450 patients who had undergone bilateral yttrium^{90} implantation in the sub-caudate regions. Hirose from Japan described a modification of Scoville's orbital undercutting and Ballantine[1] reported on experiences with bilateral stereotactic cingulotomy.

The Third International Congress of Psychosurgery was held two years later in Cambridge, England. At this time Kelly and Richardson from London[6] described their technique of "limbic leucotomy" in which cryogenic lesions were placed in the cingulum and the sub-caudate region. Bingley, Leksel *et al.* reported on stereotactic anterior capsulotomy as performed in Sweden. The enthusiasm for these international conferences was such that others were held in Madrid (1975)[10] and Boston (1978)[9].

Although the international neuroscience community seemed to have accepted psychiatric surgery other forces were denying its legitimacy. Radical groups, offspring of the "counter-culture revolution", succeeded in politicizing these surgical procedures in almost every country in which they were performed. By 1975, for example, psychiatric surgery was proscribed in Japan and severely restricted in several of the American states. The propaganda employed allegations that "psychosurgery" was synonymous with frontal lobotomy and was being used to subjugate women and minorities. At the International Congress in Boston the meeting hall was picketed. The political and social pressures became so great that most psychiatrists and neurosurgeons were reluctant to recommend and perform these procedures.

Starting in the early 1980s, however, a degree of sanity was restored regarding this issue and others of societal significance. The public was becoming increasingly well informed and in many instances it seemed to be ahead of the general psychiatric community in asking for this form of intervention when all other methods had failed.

Nevertheless, it would appear that psychiatric surgery continues to be under-utilized. In the United States, for example, statistics have been published on premature deaths (defined as deaths occurring before the age of 65) for the year 1982. Cancer was the leading cause, followed by heart disease, suicide and homicide. In that year 43,999 individuals reportedly died from suicide or homicide. Moreover, it has been estimated that on any one day in the United States there are about 45,000 individuals disabled by intractable emotional illnesses, a burden to themselves, their families and society. Since these patients become in many instances a part of the grim statistics on suicide and homicide, it would seem that, if these figures are even approximately correct and if they were more widely known, there would be a marked increase in the use of psychiatric surgery.

Another factor contributing to underutilization is the lack of publication on the part of psychiatrists and neurosurgeons. When an article appears, it is often published in an obscure journal or "proceedings". Moreover, the neurosurgeons seem to publish more frequently than the psychiatrists and it is the former that have taken the lead in assuring the viability in the United States of this type of treatment. In short, it seems incontrovertible that, as of this writing, there is a need for more surgical intervention in the treatment of intractable disorders of affect and more leadership in this direction from the psychiatrists.

A current pressing need for more clinical investigation also exists. Sub-caudate tractotomy, limbic leucotomy, anterior capsulotomy and cingulotomy all have their proponents and critics. It must be determined whether there is really a specific limbic system interruption which is best for a specific disorder of affect or: Is the limbic system such a truly reverberating circuit that the location of a lesion within it is less important than its size and shape?

An opportunity exists to assay specific neurotransmitters in the ventricular CSF and, in some instances the tissues selected for ablation, to examine the hypothesis that these substances play a crucial role in psychiatric illness. An exciting collaborative effort among psychiatrists, neurosurgeons and basic neuroscientists awaits the appearance of imaginative leaders in these areas.

Another important channel of investigation is a study of the relationship between chronic pain and depression; how do the two conditions relate to each other? Are the same neurotransmitters involved in both? Is limbic system surgery capable of returning an individual disabled by chronic pain to a productive existence?

These are all endeavors which can be mounted in the immediate future; as to the more distant future, it is highly probable that this uneasy collaboration between psychiatrists and neurosurgeons will continue. It is quite possible that the subtractions from the nervous system brought about by producing lesions in the limbic system will be replaced by the addition of neurochemicals or even of tissues known to have a favourable influence on central nervous system functioning.

Clinical investigations are well under way to explore the use of implantable pumps which will deliver accurately measured doses of psycho-pharmacological drugs via the CSF. This can result in higher central nervous system concentrations of drugs known to favourably influence psychiatric illness and, perhaps, reduce the number of unfavourable side-effects.

Most exciting are the recent attempts to treat Parkinson's Disease by tissue transplantation. It can be predicted with some degree of confidence that, as our knowledge of the crucial role that neurotransmitters play in affective illness expands these innovative methods will become more widely used.

Finally, it is to be hoped that in the present or the immediate future the ill-advised, out-moded term "psychosurgery" will be replaced by "limbic system surgery" which more accurately defines those various surgical attempts to alter the course of disorders of affect.

References

1. Ballantine Jr HT, Cassidy WL, Brodeur J, Giriunas I (1972) Frontal cingulotomy for mood disturbance. In: Hitchcock H, Laitinen L, Vaernet K (eds) Psychosurgery. Proceedings of the Second International Conference on Psychosurgery, Copenhagen, Denmark, 1970. Charles C Thomas Pub, Springfield, Ill, pp 221–229
2. Ballantine Jr HT, Bouckoms AJ, Thomas EK, Giriunas IE (1987) Treatment of Psychiatric Illness by Stereotactic Cingulotomy. Biol Psychiatry 22: 807–819
3. Egas Moniz (1936) Essai d'un traitement chirugical de certaines psychoses. Bull Acad Med (Paris) 115: 385–392
4. Freeman WL, Watts JW (1942) Psychosurgery. Charles C Thomas Pub, Springfield, Ill, p 16, pp 284–294
5. Fulton JF (1949) Functional localization in relation to frontal lobotomy. Oxford University Press, New York, pp 62–66, pp 92–93
6. Kelly D, Richardson A, Mitchell-Heggs N (1973) Technique and assessment of limbic luecotomy. In: Laitinen W, Livingston K (eds) Surgical approaches in psychiatry. Proceedings of the Third International Congress of Psychosurgery, Cambridge, England, 1972. University Park Press, Baltimore, pp 165–173
7. Knight G (1972) Bifrontal stereotaxic tractotomy in the substantis innominata: an experience of 450 cases. In: Hitchcock E, Laitinen L, Vaernet K (eds) Psychosurgery. Proceedings of the Second International Conference on Psychosurgery, Copenhagen, Denmark, 1970. Charles C Thomas Publ, Springfield, Ill, pp 269–271
8. Levin W (1961) Observations on selective leucotomy. J Neurol Neurosurg Psychiatry 24: 37–44
9. Hitchcock ER, Ballantine HT (1979) Modern concepts in psychiatric surgery. Proceedings of the Fifth World Congress of Psychiatric Surgery, Boston, Mass, U.S.A. August, 1978, Elsevier, North Holland Biomedical Press, Amsterdam
10. Neurosurgical treatment in psychiatry, pain and epilepsy (1977) Proceedings of the Fourth World Congress of Psychiatric Surgery. Madrid, Spain, 1975. University Park Press, Baltimore, pp 301–308
11. Papez JW (1937) A proposed mechanism of emotion. Arch Neurol Psychiatry 38: 725–743
12. Spiegel EH, Wycis HT (1949) Physiological and psychological results of thalamotomy. Proc R Soc Med Suppl 142: 84

Correspondence: H. Th. Ballantine Jr., M.D., Harvard Medical School, Massachusetts General Hospital, 15 Parkman Street—Suite 312, Boston, MA 02114, U.S.A.

Acta Neurochirurgica, Suppl. 44, 129–137 (1988)

Regarding the Experimental Neurophysiological Basis of Psychosurgery

A. Waltregny

Stereotactic and Functional Neurosurgical Department, University of Liège, Liège, Belgium

Summary

A survey is given, on reports in the literature and personal experiences, on the neurophysiological basis of psychosurgery. Animal experiments as well as clinical-experimental observations in human beings are reviewed, following the usual present day targets: frontal lobe, cingulum, amygdala, thalamic and hypothalamic areas and anterior internal capsule.

As a result, it has to be stated that there are no definite data resulting from animal experiments which could sustain the neurophysiological basis of psychosurgery. For psychiatric diseases animals are definitely not an adequate model. Therefore it is considered to be our neurosurgical duty to collect experimental data in human beings in a methodically proper manner and with respect to our ethical precepts.

Keywords: Psychosurgery; neurophysiology; animal experiments.

Introduction

Psychosurgery is a controversial subject and one of the main problems could be the question of its neurophysiological basis. The question of the validity and reliability of the experimental data sustaining psychosurgical procedures in human beings still seems open.

Of course the historical association of Fulton's name with the work by Moniz is quite comfortable for mind and everybody feels heartened, but what were really the actual experimental data at the time and afterwards, is a surprizing maze. To clarify this review, we will examine successively the experimental neurophysiological basis of psychosurgery according to the usual actual targets: namely the frontal white matter targets, the cingulum, the amygdala, the thalamic and hypothalamic targets and miscellaneous sites such as the corpus callosum and anterior internal capsule.

Data Related to Different Targets

1. The Frontal Poles

In the beginning of the XXth century, knowledge of the brain was growing very fast and was emerging from the philosophical interpretations which had accumulated since Babylonian, Egyptian, Greek, Roman and Arab times.

Neuronal theory was definitively established by Ramon y Cajal in 1908. Furthermore the inheritance of phrenological theory developed during the end of XIXth century, permitted the development of a new concept; the cortical localization of brain functions.

Much neurophysiological work was done in that direction, collecting considerable evidence of motor, sensory and visual specialized areas on the cortex. According to that concept, the so-called "higher functions" were also supposed to be somewhere on the cortex, but where?

Considering the phylogenetic progressive growing importance of the cerebral frontal lobes, scientists considered it as logical that modern man, the most intelligent animal by definition, would be mainly involved in this so specific anatomical particularity. Experimental researches on the frontal lobes were thus performed in the way of behavioural involvement.

Goltz in 1874, demonstrated that removal of frontal areas induces hyperactivity and sham rages in dogs, but twelve years later, Ferrier described exactly opposite symptoms; "... instead of, as before, being actively interested in their surroundings ... the animals remained apathetic, or dull, or dozed off to sleep, responding only to the sensations or impressions of the moment ...". Further similar observations were made by Rossolimo, Bechterew and Zukowski and Bianchi in different animals such as dogs, foxes and monkeys. Moreover losing the conditioned learnings was noted by Franz in cats, Kalisher in dogs and Jacobsen in monkeys.

In the historical presentation now usually referred, to, at the second International Neurological Congress in 1935, Fulton and Jacobsen presented some special observations about postoperative behaviour in two monkeys with bilateral frontal lobotomy. After both of its frontal association areas were ablated, the animal named Becky repeatedly made more errors in the delayed response test, but no longer responded to the mistakes with the "experimental neurosis" of temper tantrums. On the other hand, the animal named Lucy, had temper tantrums when after the operation it repeatedly failed the tests. Thus there is or there is not any experimental neurosis according to uncontrolled parameters. Fulton and Jacobsen concluded that the outstanding feature of the adaptation after frontal lobe ablation was the erratic and variable character of performance. However, besides these behavioural disturbances, they noticed many other somatic problems such as myoclonic jerks mainly of the head and trunk for one to two months, massive axial hypertonia, hyperreflexia, hysterical catalepsy, pyramidal signs, rage, hyperhidrosis, pupillary dilatation, temperature, piloerection, intestinal hyperactivity, loss of weight, shivers, oculomotor palsy mainly in lateral movements, increase in food intake, intestinal invaginations with obstructions, automutilations ...

It is quite incredible than in the face of such a list of big experimental problems, Moniz selected only the absence of experimental neurosis in only one out of two prefrontal lobotomized monkeys to justify his proposal to make frontal lesions in human.

So we can understand Fulton's opinion when he wrote: "At that time we were a little startled by the suggestion, for I though that Doctor Moniz envisaged a bilateral lobectomy which, though possible, would be a very formidable undertaking in a human being" (was he using the double sense of undertaking?). Nevertheless, some fifteen years later, Fulton congratulated Moniz and Lima for their imagination, perseverance and daring (he insisted)! These words of Fulton highlight cleary that the opinion of the neurophysiologists was not so clear at that time and that they would never have suggested or cautioned such operations in humans based on such experimental data in animals.

Parallel to the experimental data, clinical researchers on frontal lesions suggested at that time (Brickner 1934), that a frontal lobe damaged by a tumour is more likely to produce impairment than no lobe at all. It was thus concluded that all psychiatric symptoms are associated with aberrant frontal lobe activity and it thus became logical to consider that removal of that structure could be no less than helpful.

In the face of such a variety of controversial experimental and clinical data, Moniz correlated two facts:

1. Disappearance of frustration behaviour in only one chimpanzee with a bilateral ablation of orbitofrontal cortex (and thus ignoring that the contrary can exist!).

2. The big challenge at that time of having a unique chance, to cure or alleviate the major behavioural problems of a lot of patients in a dramatic condition of insanity and thus with no possible risk of impairment.

The first clinical results were so encouraging and convincing that the method was developed in many countries as Ballantine outlined in a previous paper (VII, 2).

The usual neurophysiological interpretation covering that procedure considers that all postoperative behavioural effects are the consequence of liberation of subcortical centers and mainly of paleothalamus according to the theory developed *a contrario* by Head, where frontal lobes have an inhibitory influence on medial thalamic nuclei.

Like others, we were seduced by the most recent tentative efforts to find correlations between symptoms and the location of lesions as suggested by Crow in Bristol. In such a protocol, a lot of gold electrodes are chronically implanted in the frontal lobes. The purpose is to find correlations between sites of coagulations and behavioural modifications. After several studies, we are now convinced that the best clinical results are not at all correlated to any particular location of the lesion, but they are well correlated with the volume of the coagulated frontal white matter, exactly as Freeman and Watts had pointed out in leucotomized patients.

Apart from this evolution of neurophysiological ideas based on the Jacksonian concept of stratified levels of functions, there were some supporters of the cortical theory of a precise functional cortical localization. So there were some investigators keen to promote topectomies involving much more localized ablations of cortical areas (see Penfield in 1947 with gyrectomy, and Pool in 1949 with removal of areas 9 and 10).

In our chronically implanted patients, we tried to see if there were any behavioural changes correlated to stimulation of the basal frontal lobes. The steroelectroencephalographic activity was recorded during special behavioural conditions like counting and manipulating some cursors up and down with the patient's own rhythm. They were no objective behavioural modifications or subjective abnormal feelings after electrical stimulation of the orbital region.

Several cooperative studies had tried to assess pre- and postoperative changes, and relate them to the location and amount of frontal lobes removed, but no clear correlations were evident and the results were fluctuating and unforeseeable with white matter lesions and with topectomies.

Thus we have no experimental data in man to demonstrate that the basofrontal cortical areas are directly implicated in behaviour.

This was the reason why some other targets were investigated and the so-called "emotional circuit of Papez" seemed a promising alternative from a neurophysiological point of view. In 1878, Broca described his "grand lobe limbique" on the medial surface of the cerebral hemisphere. In 1937, Papez described connections to hypothalamus and anterior thalamic nuclei and proposed all that as "an emotional circuit". In 1948, Yakowlev added the orbital, insular, anterior temporal cortical areas and their connections to amygdala and dorsomedial thalamus and finally in 1952, McLean added new connections to the reticular core of the brain stem. So was defined the "crown" of the mental functions (Broca had used the limbic image describing the coronal aspect of that structure around the brain stem). Subsequent anatomical and physiological researches had not given any support to the presumed olfactory function of the cingulate region, the hippocampus and some other parts of the limbic lobe, also ill-defined as "rhinencephalon" or "olfactory brain". Furthermore modern anatomical and physiological researches tend to deny form of functional unit and fractionate the limbic system into several units with quite different projections and functional significance. This is the reason why we will discuss the cingulate gyrus and amygdala in two separate sections.

2. Cingulate Gyrus

Examining the neurophysiological literature about the cingulate gyrus collected first in animal and second in human, we will review for each group the data resulting from electrical stimulations and also from surgical removals.

2.1. Stimulation of Cingulate Gyrus in Animals

In 1941, Smith reported electrical stimulations of the anterior gyrus eliciting various reactions described as expression of emotions. In exhaustive studies, Kaada and co-workers in 1949, found some tonic and clonic movements, vocalization, chewing, licking and swallowing movements, respiratory, vascular and visceral changes in monkeys, but they were also some behavioural responses. A change in behaviour termed "arrest reaction", "searching", "attention" or "arousal" can be elicited by stimulation of the anterior and less of the posterior cingulate gyrus in freely moving unanaesthetized animals. From an anthropomorphical interpretation, it is considered the facial expression changes to one of "attention" or "arousal" perhaps associated with some "surprise, bewilderment or anxiety" ... "there are slight pricking movements of the ears and quick anxious gleaming movements of the ears and head, usually to the contralateral side" ... "the animal's attention is apparently fixed on something else".

2.2. Stimulation of the Cingulate Gyrus in Humans

In 1973, Bancaud and co-workers reported on electrical stimulation of the cingulate gyrus in humans. They observed that 50 Hz repetitive stimulations at 10 volts of the anterior cingulate gyrus in conscious human subjects elicit a highly integrated type of motor behaviour. There were several simple primitive movements such as touching, leaning, rubbing, stretching or sucking in. They were executed with the fingers, lips or tongue. These movements were combined in multiple ways to yield integrated and sometimes very well adapted behavioural patterns *e.g.* sucking, palpation or nibbling. These responses were sometimes accompanied by a change in mood or level of consciousness or by autonomic phenomena. The mood change always tended towards euphoria. The level of consciousness was very rarely lowered and sometimes raised. The autonomic reactions were of several types and included mydriasis, rubefaction of the face, increase in the heart and respiration rate. Speaking and comprehension were never affected, they said.

We were working in the Bancaud's laboratory when he made these observations in 1969, and later we tried to verify and reproduce all that with a very close survey of the electrical activities not only in the cingulate gyrus but also in other surrounding structures. Our experience appears to be much more restricted. Thus, electrical stimulation of the anterior cingulate gyrus elicit huge post-discharge in the cingulum but do not modify the behaviour in any way; subjective appreciation and objective observation of motor activity and intellectual activity are strictly normal. Memory is not disturbed by this temporary disorganization of the anterior or medial cingulate region. But if we deliver similary electrical stimulation into the posterior part of the cingulum, we can see an important disorganization of vocalization, memory or motor activity. In fact they are similar to the results of stimulations of supplementary motor areas which are very closed to these cingulate regions. Thus our personal conclusions are that firing limited to the cingulate anterior gyrus induces strictly no subjective or objective changes in the behavioural status of the patient despite epileptic recruitment of all that gyrus. More complex modifications in behaviour occur only when firing spread to mesial frontal structures involving not only supplementary areas but also associative areas of the frontal poles. In our opinion, neurophysiological function of the cingulate gyrus is thus not attempted by this type of stimulation.

2.3. Ablation of the Cingulum in Animals

Earlier observations by Smith, Fulton, Ward and others have demonstrated that apprehensiveness and anxiety seem to disappear after bilateral cingulate ablation. There are behaviour changes in the direction of greater tameness and diminution of pre-operative fear and rage in response to man and in lack of "social consciousness". However cingulectomy may have the opposite effect making monkeys temporarily more aggressive or less fearful of man (Mirsky *et al.* 1957). In the same unforeseeable way, some authors like Bard 1950, Rothfield and Harman 1954, have demonstrated that bilateral lesions of the cingulate area do not alter the threshold at which rage provoking stimuli become effective. In 1955, Kennard reported cats with confused, perseverative, obsessive behaviour, a plasticity of posture and a slight increase in rage reactions. All these behaviour effects are unfortunately transient, apparently minimal and difficult to appraise as Pribram and Fulton pointed out.

2.4. Ablation of the Cingulum in Man

Remembering only the experimental reduction of anxiety, several neurosurgeons led to remove the cingulum bilaterally in agitated aggressive and overactive psychotics. Ward was the first in 1948, and many others followed (Lebeau, Scoville, Tow, Whitty ...).

An important conclusion which has emerged from all ablation studies in animal and man, is that neither unilateral or bilateral lesions of the cingulate interfere with the correct integration of basic elementary somatomotor and autonomic mechanisms, nor with functions essential for survival. Behavioural effects are present but not very striking, frequently transient and difficult to objectivate.

3. Amygdala

The other usual limbic target is the amygdala complex. Its "archaic" origin and its big relative volume in animals contributed to its being chosen in many experimental researches.

3.1. Amygdala in Animals

It is generally assumed that studies involving stimulation or ablation of the amygdala in animals have demonstrated significantly that these structures play an important role in the determination of aggressive behaviour (Kiloh *et al.* 1974), but interpretation of these phenomena is not simple because there is either aggressivity or indifference according to the experimental data.

3.1.1. Stimulation of the amygdala in animals

Electrical stimulation of the amygdala in animals produces a lot of responses including eye and head movements, visceral responses such as cardiovascular modifications, salivation, urination, ovulation, pupillary dilatation, all that considered as part of alimentary, sexual or emotional behaviour. With high level stimulations, alertness is replaced by fear and aggressivity (Egger and Flynn 1967, Storkman and Glusman 1970).

3.1.2. Ablation or lesion of the amygdala in animals

Bilateral lesions in the amygdala in animals produce marked loss of emotional reactivity. There is "indifference" toward different social stimulations: reduction of fear reactions, or defense reactions in monkeys (Plotnik 1968) and in cats (Ursin 1965), of submission in hamsters (Bunnel *et al.* 1970), of spontaneous activity in rats.

Unilateral amygdalectomy in split brain monkeys leads to maladjustment of "affective" reactions by this hemisphere in contrast to the other with an intact amygdala (Downer 1961).

Functional transitory anaesthesia of the amygdala reduced considerably autostimulation in lateral hypothalamus (Vergnes and Karli 1969).

We do not refer here to the well known experiment of Kluver and Bucy which they presented in 1937 as "the most striking behaviour changes ever produced in animals". In fact, the bilateral destruction of temporal poles involving not only rhinencephalic structures but also cortical and subcortical large areas is too massive to permit anatomo-functional correlations.

In contrast with these data, bilateral *specific* destruction of the amygdala in cats was reported by Bard and Mountcastle, to produce savageness and a considerable increase in rage reactions.

So the consequences of amygdala lesions are not unequivocal and depend mainly on the experimental conditions, and three factors seem important:

1) *type of experimental situations*: For example, a monkey with bilateral lesions in both amygdalae shows loss of fear reactions in experimental laboratory tests; but on the contrary, he has more fear when he is submitted to social interactions in a group (Thompson *et al.* 1969). In the same way, the re-introduction of such an animal into the wild is practically impossible (Kling *et al.* 1970).

2) *age of animal*: Aggressive behaviour against mice is suppressed if lesions are made in adult "killer" rats, but if lesions are made in early life, there is a significant increase of "killer" rats in the population (Karli *et al.* 1976).

3) *postoperative delay for observation*: Interspecies aggressivity can be observed as increased if observations are made more than 30 days after operation and if reactivity is enhanced by deprivation of food (Poirier and Ribadeau Dumas 1978).

Thus one work hypothesis is that the amygdala in animals is participating in a positive and negative reinforcement process. That could explain the great variability in the results of stimulation or ablations. The final result could be an algebraic sum of facilitatory or inhibitory effects, but we must recognize our precise absence of understanding of the function of the amygdala in animal behaviour.

3.2. Amygdala in Humans

3.2.1. Stimulation of the amygdala in humans

There are very few reports in the literature on electrical stimulation of the amygdala in humans. When they are reported they were performed in cases of temporal epilepsy trying to reproduce spontaneous seizures as usually performed in the work of Talairach and Bancaud. In man, anger or rage were noticed with high voltage stimulations and as part of epileptic temporal seizures.

Our personal experience in that field demonstrates that mechanical insertion of an electrode into the amygdala induces a huge electrical local post-discharge without any autonomic change. In stereoencephalographic studies, electrical stimulations elicit important post-discharges in the next hippocampus up to the cingulate gyrus on the same side but without any firing of the contralateral amygdala or hippocampus. With polygraphic testing, we can observe an immediate arrest of speech and motor activity during the entire duration of the post-discharge. There is no memory of that arrest reaction; it is a true "absence" without any automatic so-called psychomotor activity. To induce that kind of behaviour we must extend the recruitment to the temporal cortical surface with higher level stimulations.

3.2.2. Ablation of the amygdala in humans

It has been usually recognized that patients with temporal lobe seizures often show behaviour disturbances usually named "epileptoid temper" as Trimbel discussed in a previous section. In some of these patients, unilateral resection of the temporal lobe has favourably influenced the emotional and affective behaviour, sometimes even independently of the degree of seizure relief (Milner 1958, Falconer 1965). Stereotactic unilateral and bilateral amygdalotomy were reported by Narabayashi in 1961 in patients with idiocy and aggressive behaviour. Some good results were reported but often there was a recurrence after 6 to 12 months. Many authors followed this procedure and experienced up to 75% of marked reduction in aggressivity, hostility and antisocial behaviour (Balasubramaniam and Ramamurthy 1970, Vaernet and Madsen 1970). However some authors have described an initial increase of aggressiveness or emotional lability (Sawa *et al.* 1954). Finally, however there were other authors who failed to observe any obvious change in emotional behaviour after bilateral amygdalotomy (Freeman and Williams 1953, Williams 1953).

Thus from a neurophysiological point of view, bilateral destruction of amygdala has no definite effect on any particular pattern of behaviour. Nevertheless in our experience, it gives some valuable results despite the fact we are unable to assign any specific behavioural functions in amygdala.

Since Ferrier various hypotheses were proposed that a centre for memory or for behavioural organization might be located there. Experimental and clinical observations have so far failed to support the view that the amygdala might be a structure with a single, definitive and invariable function: the functional significance of these structures is still unknown.

4. Miscellaneous Targets

Some other targets like the lamella medialis of the thalamus, the corpus callosum or the anterior arm of the internal capsule were experimented on in the psychosurgical field in humans without any valuable supporting experimental animal data. These targets were discovered sometimes as fortuitous clinical observations.

Considering the fact that the lamella medialis of the thalamus has anatomic relationships with the limbic system, the orbito-frontal lobe and extrapyramidal system, it was postulated that this structure was also related to emotional integration phenomena. Bilateral lesions were made in these structures by several authors and some successes were claimed, without any complications or mental deteriorations (Poblete 1970).

Because connections between frontal lobes and anterior thalamic nuclei pass through the anterior arm of the anterior capsule, it seemed thus possible to interrupt a large number of connecting fibres with a lesion in that structure. Such a target was practically chosen by Leksell but without any neurophysiological analysis.

The anterior part of the corpus callosum is another major connection between the frontal lobes. Laitinen reported that high frequency electrical stimulation of the genu of the corpus callosum immediately abolished anxiety and tension in psychiatric patients. Simultaneously the patient experienced a feeling of inner well-being and relaxation. Subsequent permanent surgical lesions seemed to relieve the symptoms but were ineffective in cases with involutional melancholia.

All these targets were only occasionally mentioned in the literature. Clinical results seemed to be insufficiently reliable to establish any confidence into these miscellaneous surgical targets.

Conclusion

Our initial question was "are they some experimental data sustaining the validity of psychosurgical procedures?". As we have just discussed, the response is quite perplexing.

From a straightforward point of view, we must recognize that there are no definitive experimental data in animals, sustaining the neurophysiological basis of psychosurgery. I do not know any fundamental neurophysiologist who could propose transposing a valuable experimental animal target to cure any member of his own family.

But we must emphasize immediately that this is a field where animals are definitively not an adequate or adapted model. If in health, it is quite difficult to accept assimilations of animal and human behaviours, it is even more difficult to compare pathological behaviour. The animal experiments are only indications that surgery can modify in some way the animal behaviour, no more.

However, our limitations for observing and understanding animal behaviour restrain our possible interpretations in a very confined space. It is the reason why, we think that it is our neurosurgical duty to collect experimental data in humans, but of course with due regard to our ethical precepts.

References

General Articles

1. Green JR, Duisberg REH, McGrath WB (1952) Orbitofrontal lobotomy. J Neurosurg 9: 579–587
2. Holden JMC, Itil TM, Hofstatter L (1970) Prefrontal lobotomy: Stepping-stone or Pitfall? Am J Psychiatry 127: 591–598
3. Kucharski A (1984) History of frontal lobotomy in the United States 1935–1955. J Neurosurg 14: 765–772
4. Sugar O (1978) Changing attitudes toward psychosurgery. Surg Neurol 9: 331–335

Frontal Lobes

1. Bechterew W von (1911) Die Funktionen des Nervencentralsystems. G Fischer, Jena
2. Bechterew W von, Zukowski D. Cited by Messimy R
3. Bianchi L (1921) Le mécanisme du cerveau et la fonction des lobes frontaux, vol 1. Arnette, Paris
4. Brickner RM (1934) An interpretation of frontal lobe function based upon the study of a case of partial bilateral frontal lobectomy. Res Publ Assoc Res Nerv Ment Dis 13: 259–351
5. Chorover SL (1976) The pacification of the brain: from phrenology to psychosurgery. In: Morley TP (ed) Current controversies in neurosurgery. WB Saunders, Philadelphia London Toronto, pp 730–767
6. Crow HJ, Cooper R, Phillips DG (1961) Controlled multifocal frontal leucotomy for psychiatric illness. J Neurol Neurosurg Psychiat 24: 353–360
7. Crow HJ, Cooper R, Phillips DG (1963) Progressive leucotomy. Curr Psychiat Ther 3: 98–113
8. Ferrier D (1886) The functions of the brain. 2nd ed
9. Fiamberti AM (1937) Proposta di una tecnica operatoria modificata e semplificata per gli interventi alla Moniz gui lobi prefrontali in malati di mente. Rassegna di studi psichiat 26: 797
10. Franz SI (1907) On the functions of the cerebrum: the frontal lobes in relation to the production and retention of simple sensori motor habits. Arch Psychol NY 1: 1–64
11. Freeman W, Watts JW (1942) Psychosurgery, intelligence, emotion and social behaviour following prefrontal lobotomy for mental disorders. Charles C Thomas Pub, Springfield, Ill, Baltimore, Md, pp 337
12. Freeman W, Watts JW (1950) Psychosurgery in the treatment of mental disorders and intractable pain. 2nd ed. Charles C Thomas Pub, Springfield, Ill, pp 598
13. Fulton JF, Jacobsen CF (1935) The functions of the frontal lobes, a comparative study in monkeys, chimpanzees, and man. Advances in modern biology (Moscow) 4: 113–123. Abstracts Second Intern Neurol Congress, London, pp 70–71
14. Goltz F (1874) Ueber die Funktionen des Lendemarks des Hundes. Pflügers Archiv ges Physiol 8: 460–498

15. Head H (1918) Sensation and the cerebral cortex. Brain 41: pp 57 and 253
16. Hofstatter L, Smolik EA, Busch AK (1945) Prefrontal lobotomy in treatment of chronic psychoses with special reference to section of the orbital areas only. Arch Neurol Psychiat (Chicago) 53: 125–130
17. Jacobsen CF (1934) Influence of motor and premotor area lesions upon retention of acquired skilled movements, forced grasping, spasticity and vasomotor disturbance. Brain 57: 69–84
18. Kalisher cited by Feuchtwanger (1923), Die Funktionen des Stirnhirns. Springer, Berlin
19. Le Beau J (1954) Psycho-chirurgie et fonctions mentales. Masson et Cie, Paris, pp 429
20. Le Beau J, Petrie A (1952) Comparison of personality changes after, (1) prefrontal selective surgery for the relief of intractable pain and for the treatment of mental cases, (2) cingulectomy and topectomy. J Ment Sci 99: 53–61
21. McKissock W (1943) The technique of prefrontal leucotomy. J Ment Sci 89: 194–198
22. Messimy R (1939) Les effects chez le singe, de l'ablation des lobes préfrontaux. Rev Neurol 71: 1–37
23. Moniz E (1936) Tentatives opératoires dans le traitement de certaines psychoses. Masson et Cie, Paris, pp 374
24. Moniz E (1937) Prefrontal leucotomy in the treatment of mental disorders. Am J Psychiatry 93: 1379–1385
25. Morel F (1937) A propos du traitement chirurgical de la démence précoce. Schweiz Arch Neurol Pschiatr 39: 208
26. Ody F (1938) Le traitement de la démence précoce par résection du lobe préfrontal. Arch Ital de Chir 53: 321
27. Penfield W (1948) Symposium on gyrectomy. Part 1—Bilateral frontal glyrectomy and postoperative intelligence. Res Publ Ass Nerv Ment Dis 27: 519–534
28. Pool cited in Columbia—Greystone Associates (1949) Selective partial ablation of the frontal cortex. A correlative study of its effects on human psychotics subjects. FA Mettler ED NY, pp 517
29. Ramon Y Cajal S (1908) Neuron theory or reticular theory. Arch fisiol 5. Cited by Brazier MA (1959) The historical development of neurophysiology. In: Handbook of physiology Sect I, Neurophysiology. Chap I. American Physiological Society. Washington DC, pp 1–58
30. Ramon Y, Cajal S (1955) Studies on the cerebral cortex. English translation by Kraft LM. Lloyd-Luke, London
31. Rossolimo cited by Bechterew W von

Cingulum

1. Ballantine HT, Cassidy WL, Flanagan NB (1967) Stereotaxic anterior cingulectomy for neuropsychiatric illness and intractable pain. J Neurosurg 26: 488–495
2. Bard P (1950) In: Reymert ML (ed) Feelings and emotions. McGraw-Hill, New York, p 211, cited by Kaada
3. Broager B, Olesen K (1972) Psychosurgery in sixty-three cases of open cingulectomy and fourteen cases of bifrontal prehypothalamic cyrolesion. In: Hitchcock E, Laitinen L, Vaernet K (eds) Psychosurgery. Charles C Thomas, Springfield, Ill, pp 253–257
4. Broca P (1878) Anatomie comparée des circonvolutions cérébrales. Le grand lobe limbique et la scissure limbique dans la série des mammifères. Revue d'Anthropologie 1: 385–498
5. Fulton JF (1951) The physiological basis of psychosurgery. Trans Amer Phil Soc 95: 538–541
6. Kaada BR, Pribram KH, Epstein JA (1949) Respiratory and vascular responses in monkey's from temporal lobe, insula, orbital surface and cingulate gyrus. J Neurophysiol 12: 347–356
7. Kennard MA (1955) Effect of bilateral ablation of cingulate area on behaviour of cats. J Neurophysiol 18: 159–169
8. Laitinen L, Vilkki J (1972) Stereotaxic ventral anterior cingulectomy in some psychological disorders. In: Hitchcock E, Laitinen L, Vaernet K (eds) Psychosurgery. Carles C Thomas, Springfield, Ill, pp 242–252
9. Le Beau J (1954) Anterior cingulectomy in man. J Neurosurg 11: 268–276
10. McLean PD (1952) Some psychiatric implications of physiological studies on fronto-temporal portion of limbic system (visceral brain). Electroenceph Clin Neurophysiol 4: 407–418
11. Mingrino S, Shergna E (1972) Stereotaxic anterior cingulectomy in the treatment of severe behaviour disorders. In: Hitchcock E, Laitinen L, Vaernet K (eds) Psychosurgery. Charles C Thomas, Springfield, Ill, pp 258–263
12. Mirsky AF, Rosvold HE, Pribram KH (1957) Effects of cingulectomy on social behaviour in monkeys. J Neurophysiol 20: 588–601
13. Papez JW (1937) A proposed mechanism of emotion. Arch Neurol Psychiat 38: 725–743
14. Parhad MB (1953) Bilateral cingulo-tractotomy. J Neurosurg 10: 483–489
15. Pribram KH, Fulton JF (1954) An experimental critique of the effects of anterior cingulate ablations in monkey. Brain 77: 34–44
16. Rothfield L, Harman PJ (1954) On the relation of the hippocampal—fornix system to the control of rage responses in cats. J Comp Neurol 101: 265–282
17. Smith WF (1944) The effect of haemorrhage and transfusion on the ability of the rabbit to withstand reduced atmospheric pressure. Fed Proc 3: 42
18. Smith WK (1941) Vocalization and other responses elicited by excitation of the regio singularis in the monkey. Am J Physiol 133: 451–452
19. Talairach J, Bancaud J, Geier S, Bordas-Ferrer M, Bonis A, Szikla G, Rusu M (1973) The cingulate gyrus and human behaviour. Electroenceph Clin Neurophysiol 34: 45–52
20. Tow PM, Whitty CWM (1953) Personality changes after operations on the cingulate gyrus in man. J Neurol Neurosurg Psychiatry 16: 186–193
21. Ward AA (1948) The cingular gyrus: area 24. J Neurophysiol 11: 13–23
22. Whitty CWM (1955) Effects of anterior cingulectomy in man. Proc R Soc Med 48: 463–469
23. Whitty CWM, Duffield JE, Tow PM *et al* (1952) Anterior cingulectomy in the treatment of mental disease. Lancet 1: 475–481
24. Yakowlev (1948) cited by Kaada

Amygdala

1. Balasubramaniam V, Ramamurthi B (1970) Stereotaxic amygdalectomy in behaviour disorders. Confin Neurol 32: 367–373
2. Bancaud J, Talairach J, Morel P, Bresson M (1967) The amygdaloid nucleus. Clinical and electrical effects of stimulation in man. Electroenceph Clin Neurophysiol 22: 287

3. Bard P, Mountcastle VB (1949) Some forebrain mechanisms involved in expression of rage with special reference to suppression of angry behaviour. A Res Nerv and Ment Dis-Proc. 27: 362–404
4. Bunnel BN, Sodetz FJ, Shalloway DI (1970) Amygdaloid lesions and social behaviour in the golden hamster. Physiol Behav 5: 153–161
5. Downer JL (1961) Changes in visual gnostic functions and emotional behaviour following unilateral temporal pole damage in "Split-brain" monkey. Nature 191: 50–51
6. Egger MD, Flynn JP (1967) Further studies on the effects of amygdaloid stimulation and ablation on hypothalamically elicited attack behaviour in cats. In: Adey WR, Tokizane T (eds) Structure and function of the limbic system. Prog Brain Res Elsevier Publ Amsterdam 27: 165–182
7. Falconer MA (1965) The surgical treatment of temporal lobe epilepsy. Neurochirurgia 8: 161–172
8. Freeman WF, Williams J (1952) Relationship of the amygdaloid nucleus to auditory hallucinations. J Nerv Ment Dis 116: 456–462
9. Freeman WF, Williams J (1953) Hallucinations in Braille. Effects of amygdaloidectomy. Arch Neurol Psychiat 70: 630–634
10. Gibbs FA (1958) Abnormal electrical activity in the temporal regions and its relationship to abnormalities of behaviour. Res Publ Ass Res Nerv Ment Dis 36: 278–294
11. Gloor P (1960) Amygdala. In: Handbook of physiology Sect. I Neurophysiology Vol II. Amer Physiol Soc Washington DC, pp 1395–1420
12. Green JR, Duisberg REH, McGrath WB (1951) Focal epilepsy of psychomotor type. J Neurosurg 8: 157–172
13. Heath RG, Monroe RR, Mickle WA (1955) Stimulation of the amygdaloid nucleus in a schizophrenic patient. Am J Psychiat 111: 862–863
14. Heimburger RF, Whitlock CC, Kalsbeck JE (1966) Stereotaxic amygdalectomy for epilepsy with aggressive behaviour. J Am Med Ass 198: 741–745
15. Hill D (1952) EEG in episode psychotic and psycholopathic behaviour. Electroenceph Clin Neurophysiol 4: 419–442
16. Hughes JR, Schlagenhauff RE (1961) Electroclinical correlation in temporal lobe epilepsy, with emphasis on inter-areal analysis of the temporal lobe. Electroenceph Clin Neurophysiol 13: 333–339
17. Karli P (1976) Neurophysiologie du comportement. In: Kayser C (ed) Physiologie, vol 2: système nerveux, muscle. Flammarion Publ, Paris, pp 1331–1454
18. Kiloh LG, Gye RS, Rushworth RG, Bell DS, White RT (1974) Stereotactic amygdaloidotomy for aggressive behaviour. J Neurol Neurosurg Psychiatry 37: 437–444
19. King HE (1961) Psychological effects of excitation in the limbic system. In: Sheer DE (ed) Electrical stimulation of the brain. University of Texas Press, Austin USA, pp 477–486
20. Kling A (1972) Effects of amygdalectomy on social affective behaviour in non-human primates. In: Eleftherion BE (ed) The neurobiology of the amygdala. Plenum Press, New York, pp 511–536
21. Kling A, Lancaster J, Benitone J (1970) Amygdalectomy in the free-ranging vervet (cercopithecus acthiops). J Psychiat Res 7: 191–199
22. Kluver H, Bucy PC (1938) An analysis of certain effects of bilateral temporal lobectomy in the rhesus monkey, with special reference to "psychic blindness". J Psychol 5: 33–54
23. Kluver H, Bucy PC (1939) Preliminary analysis of functions of the temporal lobes in monkeys. Arch Neurol Psychiat 42: 979–1000
24. Milner B (1958) Psychological defects produced by temporal lobe excision. Res Publ Ass Res Nerv Ment Dis 36: 244–257
25. Narabayashi H (1961) Stereotaxic amygdalectomy for behaviour disorders with or without skull electroencephalographic abnormality. IInd Int Congress Neurol Surg
26. Narabayashi H, Nagoa T, Saito Y, Yoshida M, Naghata M (1963) Stereotaxic amygdalotomy for behaviour disorders. Arch Neurol Clin 9: 1–16
27. Narabayashi H, Uno M (1966) Long range results of stereotaxic amygdalotomy for behaviour. Confin neurol 27: 168–171
28. Plotnik R (1968) Changes in social behaviour of squirrel monkeys after anterior temporal lobectomy. J Comp Physiol Psychol 66: 369–377
29. Poirier J, Ribadeau Dumas J-L (1978) Le système limbique. Hoechst Publ Puteaux, pp 70
30. Pool JL (1954) Neurophysiological symposium on visceral brain of man. Neurophysiol 11: 45–63
31. Sawa M, Veki Y, Arita M, Harada I (1954) Preliminary report on the amygdaloidotomy on the psychotic patients, with interpretation of oral emotional manifestation in schizophrenics. Folia Psych Neurol Japonica 7: 309–329
32. Scoville WB (1954) The limbic lobe in man. J Neurosurg 11: 64–66
33. Storkman CIJ, Glusman M (1970) Amygdaloid modulation of hypothalamic flight in cats. J Comp Physiol Psychol 71: 365–375
34. Sweet WH, Erwin F, Mark VH (1968) The relationship of violent behaviour to focal cerebral disease. In: Garattini S, Sigg EB (eds) International symposium on the biology of aggressive behaviour. Excepta Medica, Amsterdam, Milan, pp 336–352
35. Szaba I, Rozkowska E, Kolta P (1972) Influence of contingent amygdaloid stimulation on lateral hypothalamic medial forebrain bundle self-stimulation. Physiol Behav 9: 839–849
36. Terzian H, Dalle ore G (1955) Syndrome of Kluver and Bucy reproduced in man by bilateral removal of temporal lobes. Neurology 5: 373–380
37. Thompson CI, Schwartzbaum JS, Harlow HF (1969) Development of social fear after amygdalectomy in infant rhesus monkeys. Physiol Behav 4: 249–254
38. Ursin H (1965) The effect of amygdaloid lesions on flight and defense behaviour in cats. Exp Neurol 11: 61–79
39. Vaernet K, Madsen A (1970) Stereotaxic amygdalotomy and basofrontal tractotomy in psychotics with aggressive behaviour. J Neurol Neurosurg Psychiatry 33: 858–863
40. Valladeres H, Poblete R (1956) Tratamiento quirurgico de la impulsiön y agresividad. Neurocirurgia 13: 71–77
41. Vergnes M, Karli P (1969) Effets de la stimulation de l'hypothalamus latéral, de l'amygdale et de l'hippocampe sur le comportement d'agression interspécifique rat-souris. Physiol Behav 4: 889–894
42. Williams JM (1953) Amygdaloid nucleus; clinical study of its ablation and theory as to its. Confin Neurol 13: 202–221

Miscellaneous Targets

Lamella medialis

1. Poblete M, Palestini M, Figueroa E, Gallardo R, Rojas J, Covarrubias MI, Doyharcabal Y (1970) Stereotaxic thalamotomy (Lamella medialis) in aggressive psychiatric conditions. Confin Neurol 32: 326–331

Corpus Callosum

2. Laitinen LV (1972) Stereotactic lesions in the knee of the corpus collosum in the treatment of emotional disorders. Lancet a, pp 472–475

Capsulotomy

3. Leksell L, Backlund EO (1978) Stralkirurgisk kapsulatomien oblodig operationsmetod för psykiatrisk Lä kartidningen. 75: 546–549
4. Leksell L, Herner T, Leksell D, Persson B, Lindquist C (1985) Visualization of stereotactic radiolesions by nuclear magnetic resonance. J Neurol Neurosurg Psychiatry 48: 19–20

Correspondence: A. Waltregny, M.D., Dr.Sc., Agrégé, University of Liège, Stereotactic and Functional Neurosurgical Department, Rue Louvrex, 64, B-4000 Liège, Belgium.

Acta Neurochirurgica, Suppl. 44, 138–144 (1988)

Aspects of Personality in Patients with Anxiety Disorders Undergoing Capsulotomy

P. Mindus, H. Nyman, A. Rosenquist, E. Rydin, and **B. A. Meyerson**

Departments of Psychiatry and Psychology, and Neurosurgery, Karolinska Hospital, Stockholm, Sweden

Summary

Capsulotomy is an established psychosurgical intervention for anxiety disorders. While the effectiveness of the intervention in reducing target symptoms is undisputed, the issue of negative personality changes following capsulotomy is of great concern. We studied prospectively personality traits in nine consecutive patients undergoing capsulotomy for anxiety disorder, using the Rorschach test and a personality inventory, the Karolinska Scales of Personality (KSP), administered before and one year after operation. The protocols were evaluated under blind conditions by an independent assessor who had access to no data other than the age and the sex of the patients. The Rorschach findings were used in two main comparison procedures: between the patients pre- and postoperative scores, and between that group and three reference groups. The KSP data were compared both with an age-stratified non-patient control group and with data obtained from groups of neurotic patients.

In summary, the capsulotomy patients' personalities, as expressed in their Rorschach interpretations, remained intact, and significant reductions were noted in scales reflecting anxiety and hostility. Statistically significant changes were also noted after operation in 10 of the 17 scales included in the KSP. While pathological scores were observed preoperatively in many scales, all the postoperative scores but one (Socialization) were within the normal range. Scores on the Socialization scale remained low, which is often the case in chronic patients. It is concluded that the patients displayed more normal personality features after operation than before and that adverse personality changes are not likely to occur after capsulotomy.

Keywords: Anxiety disorder; Rorschach test; personality inventory; rating; psychosurgery.

Introduction

Capsulotomy is an established psychosurgical intervention for anxiety disorders resistant to conventional treatments[1,2,3,4]. Fronto-limbic connections contained in the anterior limb of the internal capsule are intersected by way of radiofrequency heat lesions or gamma irradiation[5,6]. To date, results from over 350 capsulotomy patients are available in the literature. Different authors report remarkably similar results with improvement in two-thirds or more of those selected for treatment (for a review, see[7]). While the effectiveness of the intervention in reducing target symptoms is undisputed, it remains unclear whether any negative personality changes follow capsulotomy and, if so, their nature, frequency and severity.

In this prospective study, two fundamentally different methods, a projective test and a self-report personality inventory, were administered before, and one year after, capsulotomy to nine consecutive patients suffering from chronic and incapacitating anxiety disorders. The aim was to describe the patients personalities before operation, to relate the findings to estimates of outcome, to examine pre- and postoperative differences indicative of personality changes, and to compare the findings with relevant data in the literature. This has not been done before.

Patients

All patients operated upon with capsulotomy at our department during the study period of twelve months entered into the study. The probands were four female and five male consecutive patients, all satisfying the DSM III[8] criteria for Anxiety Disorders. No patient had an Axis II diagnosis of Personality Disorder. Details of the selection procedure is given in[2]. Their mean duration of illness exceeded 15 (range 6–40) years, reflecting the chronicity of their conditions. All patients had repeatedly been in hospital for their disorders, which had proved resistant to available conventional forms of treatment. Demographic and clinical data are given in Table 1. The study was approved by an Ethics Committee, and all patients gave informed consent.

Methods

All patients were operated on by the same neurosurgeon (B A M) according to a technique described in detail elsewhere[1,2]. The target area in the internal capsule was localized with stereotactic CT. Bipolar electrodes introduced stereotactically into the target area were used to produce radiofrequency lesions in the capsular white matter. The size of the lesion has been estimated to be 8 by 18 mm[1].

Psychiatric Evaluation

All patients were evaluated by the same psychiatrist (P M). Clinical morbidity before and after operation was rated on the Comprehensive Psychopathological Rating Scale (CPRS). This scale has been shown to be sensitive to symptom changes induced by psychiatric treatment, and to have high validity and intra- and inter-assessor reliabilities. (For a description of the scale and validity and reliability data, see[9]). The rating sessions were recorded on audiotapes used for the evaluation of both intra- and inter-assessor reliabilities between PM and a second assesssor, HN (who is independent of the psychosurgery team). In addition to this symptom-oriented measure, the patient's level of social functioning before and after the operation was rated on the Axis V of the DSM III[8]. The Pippard postoperative rating scale[10] was chosen to estimate clinical outcome, since that scale has been used widely in the psychosurgical literature. (Details of the evaluation procedure will be published elsewhere.)

Personality Traits

Several authors have employed inventories to study personality change in relation to various forms of psychosurgery, including capsulotomy[2, 3, 11] (for reviews see[12, 13]). Although inventories may be critized on many grounds, such as fakability ("faking good") of response or influence of response style, it has been concluded that inventories represent "the best general approach currently available to measuring personality characteristics"[14]. It may be assumed that personality inventories reflect the patient's own, mostly conscious, image of himself. Projective tests, on the other hand, are believed to reflect both conscious and unconscious aspects of personality, and may be regarded as supplementary to personality inventories. For the purpose of the present study, both types of tests were included.

The Rorschach Test

The Rorschach test is probably the most well-known projective method. Several scoring systems for the test have been developed and shown to have high interjudge reliability and substantial predictive validity (for reviews, see[15, 16]). The Rorschach test was administered by a clinical psychologist (HN) two weeks before and again one year after operation. The administration and initial scoring procedure were performed according to clinical routine as described in detail elsewhere[17, 18]. The protocols were coded with information on the age and the sex of each patient and later evaluated under blind conditions by a second independent psychologist (AR).

The Rorschach material may be analyzed in several ways. Here we report only the results on the following scales: The Friedman's indices of developmental level (DL) and of integration (IL), the Elizur's Anxiety scoring, the Elizur's Hostility scoring, and the Piotrowski Organic Signs (for description of the scales, the scoring procedures and reliability data, see[15, 19]. The DL and the IL are computed from the Friedman's Developmental Level scoring, an estimate of the degree of primitive and psychotic functioning vs mature and neurotic functioning. The AL scale is assumed to reflect anxiety proneness, and the HL score is a measure of the hostility expressed the Rorschach responses. The Rorschach findings were used in two main comparison procedures: between the capsulotomy patients' pre- and postoperative scores, and between that group and three reference groups. Each capsulotomy patient was compared with regard to the Rorschach scales with an age- and sex-matched reference subject drawn from a sample of non-surgery patients suffering from obsessive-compulsive disorder (OCD) (described by Rydin *et al.*[16]). The protocols of the nine matched reference subjects were re-scored independently and blindly by one of us (AR). The mean scores of the capsulotomy group on the DL, AL and HL scales were compared with the corresponding scores reported by Singer and Larsson[19] on 20 normal and 25 neurotic and by Elizur[15] on 22 normal and 22 neurotic subjects.

The Karolinska Scales of Personality (KSP)

The KSP is a well researched self-report personality inventory developed by Schalling *et al.*[14]. It contains scales reflecting aspects of *e.g.* anxiety proneness, impulsivity, aggression and hostility. The items were selected on the basis of theoretical considerations or drawn from published scales. The KSP personality inventory has been used in a large number of studies both on healthy subjects and on patients with various psychiatric and psychosomatic disorders. Several studies have shown the scales to differentiate between patients suffering from anxiety disorders, psychosomatic disorders and controls. Reviews of these studies have been given by Schalling[14, 20].

The KSP is a self-report questionnaire with 17 scales consisting of 135 items, each with four response alternatives from "Does not apply at all" to "Applies completely". The raw scores may be transformed into normative T-scores (mean = 50, SD = 10) based on data from an age-stratified (range 23–65 years) non-patient control group of 228 males and 240 females randomly collected from a suburb near Stockholm[21]. Four scales concern anxiety proneness: 1. Somatic Anxiety, 2. Psychic Anxiety, 3. Muscular Tension, 4. The Multi-Component Anxiety scale, which is the mean of the three anxiety scales. Three scales are related to Extraversion as conceptualized in the Eysenck Personality Inventory (EPI)[22]: 5. Detachment, 6. Impulsiveness and 7. Monotony Avoidance. Other scales are: 8. Psychasthenia, Social Desirability (this scale was excluded from the present study because of lack of normative T-scores), and 9. Socialization.

An additional number of scales describe aspects of aggression and of hostility. All scales are listed in the Table 2 (For details, see[14, 20]).

Results

Clinical Results

There were no surgical complications.

The intra- and the inter-assessor reliabilities of the CPRS scores were both high (product moment correlations 0.96/0.92, and 0.96/0.87 for pre- and postoperative ratings respectively). A one way analysis of variance of mean summed CPRS scores rated before and after operation showed a highly significant postoperative reduction ($F(3.27) = 17.67$ $p < 0.001$). Also, when comparing the patients' CPRS target symptom quotient (this measure was computed by dividing the patient's summed CPRS score on target items with the number of such items), a statistically highly significant pre- and postoperative difference was obtained ($t = 7.15$ $p < 0.001$). The subjects' social functioning had improved from a mean score of 6 ("very poor") to 4 ("fair") one year after operation. The effect of the intervention in reducing target symptoms as rated on the Pippard scale was satisfactory in all but two patients. The results are summarized in Table 1.

Table 1. *Demographic and Clinical Data from Nine Patients with Anxiety Disorder Undergoing Capsulotomy*

Case	Sex	Diagnosis	Duration of illness (yr.)	Age at operation	Preoperative score		Postoperative score		
					Target	Axis V	Target	Axis V	Pippard
1. YB	F	300.21	10	36	2.4	5	0.5	4	B
2. PP	M	300.30	10	39	1.5	6	0	4	A
3. TP	M	300.02	30	54	2.0	7	0.3	4	A
4. BS	M	300.21	15	29	2.1	6	0	4	A
5. BE	F	300.30	6	32	2.2	6	2.1	6	D
6. ML	F	300.30	23	53	2.3	6	0.7	5	C
7. RW	M	300.30	14	37	2.4	6	0	5	A
8. BO	M	300.30	22	42	2.7	6	0	3	A
9. MO	F	300.02	9	39	2.0	7	0.3	5	B
			$\bar{x} = 15.4$	$\bar{x} = 40.1$	$\bar{x} = 2.17$	$\bar{x} = 6.1$	$\bar{x} = 0.43$	$\bar{x} = 4.4$	

The CPRS target symptom quotient was computed by dividing the patient's CPRS scores on target items with the number of target items. Axis V is a scale of the level of functioning included in the DSM III, Pippard refers to the Pippard's postoperative rating scale.

Table 2. *Scores on Four Rorschach Scales Recorded in Capsulotomy Patients and in Three Comparison Groups*

Subjects	Developmental level			Integration level			Anxiety level		Hostility level	
	Pre	Post		Pre	Post		Pre	Post	Pre	Post
Capsulotomy group M (SD)	3.78 (0.46)	3.66 (0.37)	NS	21.96 (14.8)	19.50 (16.0)	NS	29.44 (19.8)	14.11[d] (8.8)	29.56 (9.6)	18.66[d] (14.7)
Reference[a] group of OCD M (SD) patients	3.16 (0.37)			9.96 (7.8)			13.89 (15.6)		18.22 (13.9)	
Neurotics[b]	3.49			17.67			12.50		5.60	
Normals[c]	3.59			23.40			5.20		1.30	

[a] Data from Ref. 16.
[b] Data from Ref. 19.
[c] Data from Ref. 15.
M = mean, (SD) = Standard deviation of the mean.
NS = non-significant.
[d] = $p < 0.05$.

The Rorschach Findings

All capsulotomy patients appeared to have reached the level of neurotic functioning as determined from the Rorschach material (recorded preoperatively). In no case were there indications of psychosis or of Borderline Personality Disorder. Before operation, the patients' interpretations of the Rorschach tables often contained responses indicative of fear of bodily harm or of high levels of anxiety (*e.g.* "no head, the head is missing, the head is very small, strange heads"). After operation the patients responses contained similar notions, but generally brief and undramatically expressed. So-called popular interpretations, which are often given by non-patient subjects, were clearly more common after operation than before. (A full account of the Rorschach findings will be published elsewhere.)

After operation, there were statistically significant reductions in the scores on the AL and the HL scales, whereas the DL and the IL scores remained unchanged. Generally, the capsulotomy patients postoperative scores approached those of the reference groups. There were no statistically significant differences in the pre- and postoperative mean scores on the Piotrowski Organic Signs scale. The results are summarized in Table 2.

For comparative purposes, the capsulotomy patients preoperative scores on the Rorschach scales were related to those of four other groups. Their mean DL score was significantly higher ($t = 3.9$ $p < 0.001$) than that of the group of non-surgery obsessive compulsive disorder (OCD) patients matched for age and sex, and also higher than those reported in the literature for groups of neurotics and normals. The IL score of the capsulotomy group was also significantly higher ($t = 2.15$ $p < 0.05$), than that of the reference group. On the AL scale the score of the operated group was numerically higher than that of the other patient groups, approaching the $p < 0.05$ level of statistical significance. When comparing the capsulotomy patients preoperative scores on the HL scale with that of the non-surgery OCD patients, there was a trend ($t = 2.05$ $p < 0.10$) to higher scores among the former, *i.e.* the operated group expressed more hostility in their Rorschach interpretations.

The Karolinska Scales of Personality

The results are shown in Table 3. The capsulotomy patients preoperative scores were abnormal on all scales reflecting anxiety, and on the scales Psychasthenia and Socialization. Borderline scores were noted on the Inhibition of Aggression and on the Irritability scales. As mentioned in the Methods section, the score on the MCA scale is computed as the mean of the scores on the three anxiety scales. Before operation, severely pathological scores were obtained, reflecting high levels of anxiety. One year after operation, a statistically significant reduction was found, with MCA scores within the normal range, see Fig. 1. Furthermore, statistically significant reductions were noted postoperatively on a majority of the KSP scales. As seen in Table 2, a normalization was found in the scores on all scales with one exception, the Socialization scale, which remained low.

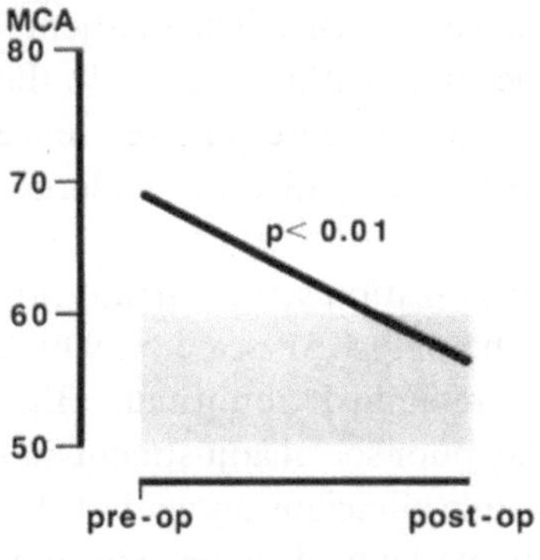

Fig. 1

Table 3. *Scores on the Karolinska Scales of Personality Recorded Before and One Year After Capsulotomy*

KSP scale	Preop.		Postop.		t (df = 7)
	Mean	SD	Mean	SD	
Somatic anxiety	66.3	14.1	57.6	8.5	2.36*
Psychic anxiety	67.1	10.4	54.7	12.9	2.77*
Muscular tens.	70.3	20.1	54.7	12.9	2.92*
MCA	68.0	13.2	57.4	10.7	3.69**
Impulsivity	44.7	14.5	54.3	14.9	—2.65*
Monotony avoid	48.9	7.1	56.1	12.5	—2.65*
Detachment	47.0	10.0	47.6	11.3	—0.15
Psychasthenia	73.6	12.5	58.6	11.1	3.85**
Socialization	36.0	12.3	38.7	10.6	—1.38
Indirect aggr.	56.0	14.0	49.7	9.6	2.20*
Verbal aggr.	48.0	14.2	51.4	9.3	—0.94
Irritability	59.7	9.1	50.8	7.4	2.85*
Suspicion	56.1	10.2	55.7	14.3	0.10
Guilt	57.3	8.0	52.3	12.0	1.22
Aggression	54.3	11.2	50.6	6.5	1.58
Hostility	56.9	3.5	54.2	11.6	0.79
Inhibition of aggr.	60.0	6.9	50.3	9.8	2.81*

Raw scores were transformed into normative T-scores with M = 50, SD = ± 10.

* = $p < 0.05$.

** $p < 0.01$.

Two-tailed t-test between pre- and postoperative scores, correlated data.

Discussion

In this prospective study, personality changes related to capsulotomy were examined under blind conditions using a projective test and a personality inventory. The nine subjects were all the consecutive patients undergoing capsulotomy at our department during the study period. We report data from this limited number of subjects for three main reasons. Firstly, the issue of personality changes related to capsulotomy, if any, is of much concern. Secondly, and perhaps a consequence of the aforementioned, only few patients undergo psychosurgical intervention. Thirdly, although there is evidence that the level of anxiety affects personality measures[23], remarkably few studies have directly compared such measures in anxiety patients before and after treatment.

One year after operation, clinically and statistically significant reductions in target symptoms were recorded in seven of the nine patients. The results agree with those reported from larger series[1, 2, 3, 4], in which two-thirds or more of the patients were reported to benefit from the operation. The patients ego organization as reflected in their Rorschach responses remained intact. So-called normal responses, which are given by healthy subjects exposed to the Rorschach tables, were clearly more common after operation than before, and significant reductions were noted in anxiety and hostility measures. A return to normal was also observed in the postoperative scores on a majority of the scales contained in the personality inventory. In summary, the patients displayed more normal personality features after operation than before. Memory effects on the postoperative scores of both the Rorschach and the KSP cannot be ruled out completely. It appears plausible, however, that any such effect should reduce the pre- and postoperative differences observed.

The standardized scoring system of the Rorschach test used in the study has been shown to have high interjudge reliability and substantial predictive validity[15, 16]. The protocols were evaluated under blind conditions by an independent assessor, who had access to no other data than the age and the sex of the patients. It is therefore unlikely that observer bias played an important role in the evaluation procedure.

No statistically significant differences were obtained between the pre- and postoperative DL and IL scores. This finding indicates that the patients ego organizations remained unchanged by the operation, and that there were no signs of psychopathy. The differences in scores on the DL and IL scales between the capsulotomy and the non-operated group of matched OCD patients can be attributed to selection effects. To become eligible for capsulotomy, a patient must display mature personality and considerable ego strength. No such selection had been made in the OCD patients, who were so severely and acutely ill that they had to be admitted to hospital. The differences in scores between the capsulotomy group and those reported by Singer and Larsson[19] could be due, in part, to age differences: their patients mean age was 19.9 compared to 40.1 year in our group. Younger subjects may be expected to score lower on the DL and the IL scales. The capsulotomy patients preoperative scores on the AL scale, assumed to reflect anxiety proneness, were very high. This is an expected finding which is in line with clinical experience and may reflect the selection procedure. As for the HL scale, the operated group expressed more hostility in their preoperative Rorschach responses than at follow-up. At that time their scores approached those of both the reference group of non-operated OCD patients and reference groups reported in the literature. It must be pointed out, however, that the control data available in the literature for comparison are afflicted with many of the shortcomings described by Goldfried *et al.*[15], *e.g.* small groups and insufficient background information. The absence of pre- and postoperative differences on the Piotrowski Organic Sign scale was expected. It is improbable that minimal lesions in the internal capsule will result in dysfunctions observable with such a test for organic signs. That contention is supported by the fact, that several previous studies have failed to show intellectual deficits after capsulotomy[2, 3, 11, 24].

The KSP scores were computed independently and blindly. The KSP, like other inventories, may be criticised on several grounds such as fakability ("faking good"), or influence of response style. The inventory has particular psychometric properties which should reduce such effects, including a so-called "lie-scale", Social Desirability (SD). The scores on the SD scale did not reveal significant effects of the factors mentioned. Furthermore, it is unlikely that such factors would be differently influential on the pre- and postoperative scores respectively.

Before operation, abnormal KSP scores were noted on the scales directly related to anxiety, an expected finding in view of the selection criteria. After operation, significant reduction were noted in these scales with scores within the normal range, according to the norms for the KSP. Similar findings have been reported after capsulotomy by other authors using identical or similar measures. Thus, Bingley *et al.*[2] reported significant reductions in variables such as anxiety, depression, obsessive thinking, neuroticism and introversion. Rylander[3] noted significant reductions in scores on the four anxiety scales included in the KSP in patients undergoing gamma capsulotomy. In a long-term follow-up study, Mindus *et al.* (unpublished data) gave the KSP to seven consecutive patients who had undergone gamma capsulotomy seven years earlier. Their patients' scores on the anxiety scales were within or close to the normal range, and lower than the scores of a reference group of 16 in-patients with anxiety disorder undergoing a drug trial.

The findings on the Socialization scale warrant further comment. This was the only KSP scale on which the postoperative scores remained abnormal. High scorers on this scale display poor social adjustment and general dissatisfaction. While clinicians agree that this is often the case in chronic patients, disagreement prevails on the causal or effect issue.

Reich *et al.*[25] gave similar personality scales to anxiety patients undergoing a drug trial. Inspection of their data reveals that both male and female patients displayed significantly lower socialization scores than controls both in the ill and the recovered state. Thus, both in their study and in the present one the findings support the common clinical impression that patients with chronic disorders exhibit difficulties in their social adjustment.

As seen in Table 2, very high preoperative scores were obtained on the Psychasthenia scale. High scores in this measure display uneasiness and lack of energy. It can be debated, whether this is a cause, an effect or simply an epiphenomenon of a severe, chronic disorder. Ultimately, only prospective studies can answer that question. At present, there is only one such report in the literature, a study by the Swedish researchers Nyström and Lindegård[26]. They followed-up a population sample consisting of 3,019 male subjects for six years. Indices of premorbid psychasthenia (according to Sjöbring) were strongly associated with mental disorders in general, and had a close connection to anxiety states in particular. Their findings could be interpreted as meaning that psychasthenic individuals run a high risk of later developing mental disorders. We do not yet know, however, whether the effect is pathogenic, *e.g.* by way of a vulnerability factor, or merely a pathoplastic influence, an effect on the clinical form of the disturbance.

In a previous study, Reich *et al.*[23] found that anxiety patients, who had improved with medication, displayed significant changes compared to baseline, *i.e.* towards normalization. They showed *e.g.* increased emotional strength and extraversion, and decreased interpersonal dependency. Our own findings support those reported by Reich *et al.*[23]. Although different personality estimates were used in the two studies on different patient populations, it may be concluded that anxiety patients display more normal personality features when recovered, irrespective of the type of treatment.

Clinical, theoretical and research implications of the findings include the following. The opinion that psychosurgery renders "the person less able, more simple or childlike, ultimately making the person susceptible to manipulation and control"[27], does not apply to capsulotomy. On the contrary, significantly more normal personality features are noted after capsulotomy than before[2, 3, 24]. Personality inventories, such as the KSP, should be used for screening purposes. The KSP is easy to administer and evaluate and is provided with norms. Should pathological personality traits be revealed by clinical interview and personality inventory, the Rorschach test may prove useful in elucidating more accurately the type of personality disturbance. Since the procedure involves both time and expertise it may not be suitable for routine screening purposes, however.

It is of theoretical interest whether anxiety disorder is associated with a distortion of the patient's cognitive assessment of his functioning, or with a personality dysfunction that is adequately reported, or both. The level of anxiety, *i.e.* state anxiety, affects personality measures. This fact that must be borne in mind both when interpreting published reports, and when planning future studies on the relationship between personality and anxiety disorder.

Acknowledgements

Supported by the Swedish Medical Research Council (4545, 5454), the Söderström-Königska Foundation, and the Salus Foundation.

References

1. Herner T (1961) Treatment of mental disorders with frontal stereotactic thermo-lesions. A follow-up study of 116 cases. Acta Psychiatr Neurol Scand [Suppl] 158: 36
2. Bingley T, Leksell L, Meyerson BA, Rylander G (1977) Long term results of stereotactic capsulotomy in chronic obsessive-compulsive neurosis. In: Sweet H, Obrador S, Martin-Rodriguez JG (eds) Neurosurgical treatment in psychiatry, pain and epilepsy. University Park Press, Baltimore, pp 287–289
3. Rylander G (1979) Stereotactic radiosurgery in anxiety and obsessive-compulsive states: psychiatric aspects. In: Hitchcock ER, Ballantine Jr HT, Meyerson BA (eds) Modern concepts in psychiatric surgery. Elsevier, North-Holland Biomedical Press, Amsterdam New York Oxford, pp 235–240
4. Burzaco J (1981) Stereotactic surgery in the treatment of obsessive-compulsive neurosis. In: Perris C *et al* (eds) Biological psychiatry. Elsevier, North-Holland Biomedical Press, Amsterdam New York Oxford, pp 1103–1109
5. Leksell L, Backlund EO (1979) Stereotactic gammacapsulotomy. In: Hitchcock ER, Ballantine Jr HT, Meyerson BA (eds) Modern concepts in psychiatric surgery. Elsevier, North-Holland Biomedical Press, Amsterdam New York Oxford, pp 213–216
6. Leksell L (1983) Stereotactic radiosurgery. J Neurol Neurosurg Psychiatry 46: 797–803
7. Meyerson BA, Mindus P (1988) Capsulotomy as treatment of anxiety disorders. In: Dade Lunsford (ed) Modern stereotactic neurosurgery, Martinus Nijhoff Publ, Boston, pp 353–364
8. Diagnostic and statistical manual of mental disorders. DSM III (1980) American Psychiatric Association. Task force on nomenclature and statistics. Washington
9. Åsberg M, Montgomery S, Perris C, Schalling D, Sedvall G (1978) CPRS—The Comprehensive Psychopathological Rating Scale. Acta Psychiatr Scand [Suppl] 271: 5–27
10. Pippard J (1955) Rostral leucotomy: a report on 240 cases personally followed-up after $1\,^1/_2$ to 5 years. J Ment Sci 101: 756–777
11. Fodstad H, Strandman E, Karlsson B, West KA (1982) Treat-

ment of chronic obsessive compulsive states with stereotactic anterior capsulotomy or cingulotomy. Acta Neurochir (Wien) 62: 1–23
12. Kelly D (1980) Anxiety and emotions. Psychological Basis and Treatment. Charles C Thomas, Springfield, Ill
13. Corkin S (1980) A Prospective Study of Cingulotomy. In: Vallenstein ES (ed) The psychosurgery debate. Scientific, legal and ethical perspectives. SH Freeman and Co, San Francisco
14. Schalling D (1986) The development of the KSP inventory. In: Report from the research program individual development and adjustment, Dept of Psychology, University of Stockholm. No 64
15. Goldfried MR, Stricker G, Weiner IB (1971) Rorschach handbook of clinical and research applications. New Jersey, Prentice Hall
16. Rydin E, Thorén P, Åsberg M (1987) Rorschach prediction of response to clomipramine treatment in chronic obsessive compulsive disorder. Unpubl manuscript
17. Klopfer B, Ainsworth MD, Klopfer WG, Holt RR (1954) Developments in the Rorschach technique, vol 1. Harcourt, Brace & Company, New York
18. Beck SJ, Beck AG, Levitt ET, Molish HB (1961) Rorschach's test. I. Basic processes. Grune & Stratton, New York
19. Singer MT, Larsson DG (1981) Borderline Personality and the Rorschach Test. Arch Gen Psychiatry vol 38
20. Schalling D, Åsberg M, Edman G, Oreland L (1987) Markers for vulnerability to psychopathology: temperament traits associated with platelet MAO activity. Acta Psychiatr Scand 76: 172–182
21. Bergman H, Bergman I, Engelbrektson K, Holm L, Johannesson K, Lindberg S (1982) Psykologhandbok. Del 1. (Psychological Manual Part 1). Magnus Huss Klinik, Stockholm
22. Eysenck HJ (1981) A model of personality. Springer, New York
23. Reich J, Noyes Jr R, Corywell W, O'Gorman T (1986) The effect of state anxiety on personality measurement. Am J Psychiatry 143: 760–763
24. Kullberg G (1977) Differences in effect of capsulotomy and cingulotomy. In: Sweet WH, Obrador S, Martin-Rodriguez JG (eds) Neurosurgical treatment in psychiatry, pain and epilepsy. University Park Press, Baltimore
25. Reich J, Noyes Jr R, Hirschfeld R, Corywell W, O'Gorman T (1987) State and personality in depressed and panic patients. Am J Psychiatry 144: 181–187
26. Nyström S, Lindegård B (1975) Predisposition for mental syndromes: a study comparing predisposition for depression, neurasthenia and anxiety state. Acta Psychiat Scand 51: 69–76
27. Breggin PR (1980) Brain-disabling therapies. In: Vallenstein ES (ed) The psychosurgery debate. Scientific, legal and ethical perspectives. SH Freeman and Co, San Franciso

Correspondence: P. Mindus, M.D., Department of Neurosurgery, Karolinska Hospital, S-10401 Stockholm, Sweden.

Acta Neurochirurgica, Suppl. 44, 145–151 (1988)

Posteromedial Hypothalamotomy in the Treatment of Violent, Aggressive Behaviour

K. Sano* and **Y. Mayanagi****

Departments of Neurosurgery, Teikyo University and University of Tokyo*, Tokyo Metropolitan Police Hospital and University of Tokyo**, Japan

Summary

Although emotion in the human is largely modified by the frontal association areas (software) and may better be called affect, it is still very much influenced by the balance of the ergotropic and the trophotropic circuits in the prosencephalon (hardware) especially in patients with organic brain lesions. Violent, aggressive, restless behaviours or rage can be regarded as an unbalanced state of these two circuits with dominance of the ergotropic circuit. In order to restore the balance of these two circuits, small stereotactic lesions were made in the ergotropic portion of the posterior hypothalamus (posteromedial hypothalamotomy) with good results in the follow-up of 10–25 years. Postoperatively there was no disturbance in endocrine activities and growth.

Keywords: Hypothalamus; hypothalamotomy; aggressive behaviour; emotion.

Introduction

Personality is the integrated activity of all the reaction-tendencies of the individual. It is, therefore, the person as he (or she) is known to his (or her) friends or companions. We may regard personality as the total integrated expression of the various neural levels of which the individual is constructed—the lowest or the brain stem-spinal cord level, the limbic system (including the diencephalon) level, the neocortex level and the level of the frontal association areas.

Figure 1 illustrates various functions at various levels of the central nervous system[13]. It also shows the difference between the emotional and affective behaviours.

Driven simply by instincts, all animals—including humans—would live as vigorously as possible. Limited food, limited material and limited sex would lead inevitably to the struggle for existence. "Emotion" is essential to survive this struggle. Animals feel pleasure when instincts are satisfied and displeasure when they are not. The displeasure may be expressed as anger or rage, fear or horror. We call these sensations "emotion". The limbic system and the brain stem, the brain's "hardware" in computer terminology, most probably constitute the neural basis of emotion.

Emotion in this sense is "dry". Two animals may fight with the most intense rage, yet the fight seldom leads to intraspecies killing. The loser will flee, showing fear, and the fight is over.

The human being has no pure emotion, except in infancy. Human emotion is always modified by the frontal association areas which are most conspicuously developed in Homo sapiens and may be compared to the "software". This software-modified emotion in the human is "wet" and may better be called "affect". The affect is composed of gladness, happiness, sorrow, anxiety, apprehension, jealousy, grudge and so forth. Contrary to the rage of animals, aggression in the human may lead to murder, because of wetness of affect. If functions of the software are decreased in brain-damaged men, their aggressive behaviours are emotional and dry rather than affective and wet.

Aggressive behaviour may be defined as behaviour

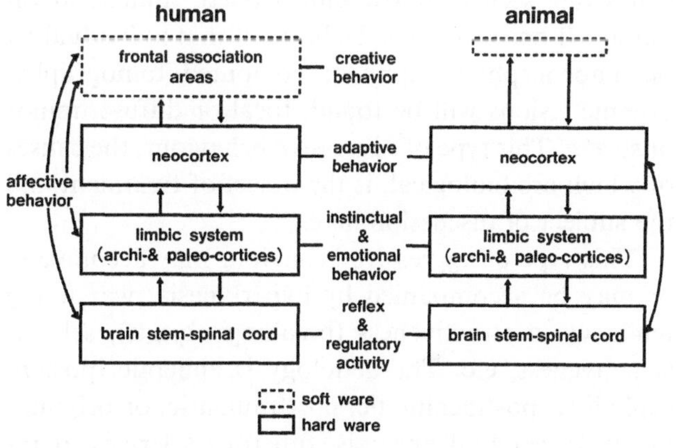

Fig. 1. Functions of various levels of the central nervous system (modified from Sano[13])

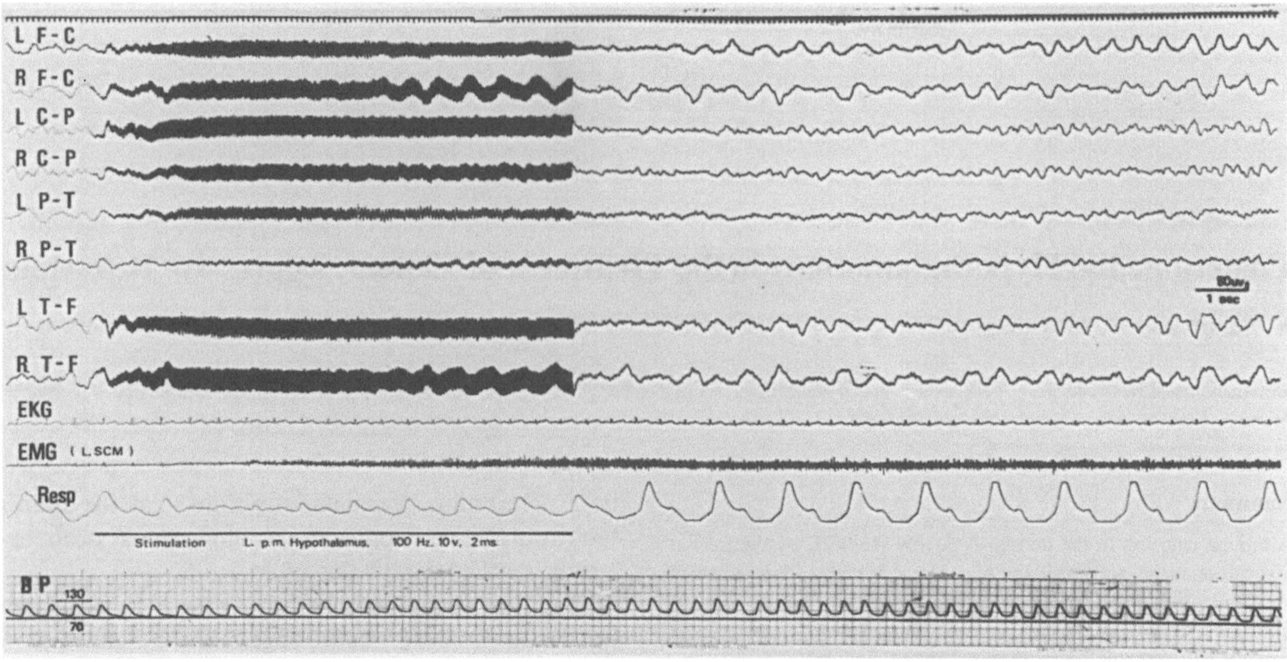

Fig. 2. EEG and polygraph recordings to show responses caused by high frequency stimulation of the left posterior hypothalamus under general anaesthesia. F = frontal, C = central, P = parietal, T = temporal, L = left, R = right, EKG = electrocardiogram, EMG = electromyogram recorded from the left sternocleidomastoid muscle, Resp = respiratory rhythm recorded on the chest, BP = blood pressure from the right radial artery

that leads to damage or destruction of some objective and is not necessarily considered to be provoked in the usual sense. If one calls the inner feeling behind the aggressive behaviour "anger", it is really a short madness (Ira furor brevis est) as Horace [Quintus Horatius Flaccus (65-8 BC)] put it in his Epistles. This anger is not the anger, the causes of which are understandable and about which Thomas Fuller (1608–1661) remarked in his "The Holy State and the Profane State", that "Anger is one of the sinews of the soul; he that wants it hath a maimed mind". The manifested aggressive behaviour seems almost unprovoked, almost involuntary. If one examines the brain of that individual by pneumoencephalography or computed tomography, organic lesions will be found, focal or diffuse, minor or severe. This type of aggressive behaviour, the causes of which are biological, is the object of treatment and the subject of discussion here.

This type of aggressive behaviour may accompany or may be accompanied by hyperkinesia, wandering tendency, destructiveness (breaking objects), self-destructiveness, etc. The aetiology is epileptic, postencephalitic, postmeningitic, posttraumatic, or originating from cerebral agenesis, tuberous sclerosis, paranatal damage, etc.

Animal experiments by Hess and others[3, 5, 6] and by us[11] showed that stimulation of the posterior portion of the hypothalamus or destruction of the anterior portion of the hypothalamus made the animal hyperactive and aggressive while stimulation of the anterior portion of the hypothalamus or destruction of the posterior portion of the hypothalamus made the animal placid and calm. Based on these experimental data, in 1962, Sano started to make stereotactic lesions in the posteromedial part of the hypothalamus where most marked signs of sympathetic discharge were elicited upon electrical stimulation, in order to calm down or correct patients with violent, aggressive behaviour. He called this procedure "posteromedial hypothalamotomy"[10, 11, 13, 14, 15, 17]. Later this procedure was also found to be effective in alleviating some types of intractable pain[12, 16].

Technique

Endotracheal general anaesthesia is used. A frontal burr-hole is made and a fine catheter is inserted into the anterior horn of the lateral ventricle, placing the tip around the foramen of Monro. A ventriculography is performed with air or a small amount of positive contrast medium. If the tip of the ventricle tube is easily introduced into the third ventricle, a small amount of CSF is taken repeatedly for analysis. Setting the anterior and posterior commissures as a guide for measurement, a fine concentric bipolar electrode of about 0.8 mm or 1 mm in outer diameter with an interpolar distance of 0.5 mm is

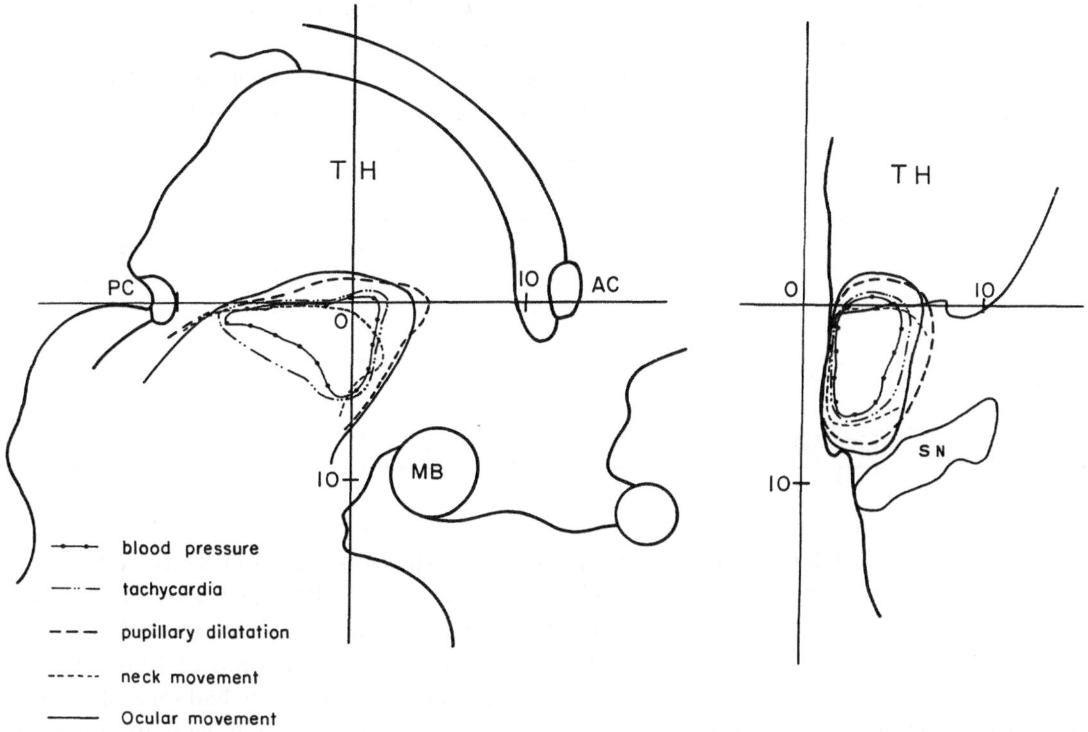

Fig. 3. Summary of autonomic and somatomotor responses obtained by high frequency stimulation of the posterior hypothalamus. Blood pressure = elevation of blood pressure upon stimulation, AC = anterior commissure, MB = mammillary body, O = midcommissural point, PC = posterior commissure, SN = substantia nigra, TH = thalamus (from Sano *et al*[15])

stereotactically inserted through a frontal burr-hole into the target point of the posterior hypothalamus.

The target point for behavioural disturbances is in most cases 2 mm below the midcommissural point and 2 mm lateral to the third ventricle wall. EEG, ECG, EMG of the neck muscles, blood pressure and respiration are monitored. Electrical stimulation is given with parameters of 10–100 Hz, 10–20 volts with square pulses of 1–2 msec in pulse duration.

Responses caused by high frequency stimulation of the target point can be summarized as follows:

(1) Autonomic respones, mainly sympathetic; Elevation of blood pressure; Increase of pulse rate; Respiratory suppression, followed by hyperpnoea or tachypnoea; Mydriasis; Flushing of the face.

(2) Somatomotor responses; Rotatory movement of the eye ball on the stimulated side; downward or inward; Lateral flexion of the neck toward the stimulated side.

(3) EEG responses; Diffuse slow wave bursts of 2–3 Hz (under light general anaesthesia) continuing for several minutes even after cessation of stimulation; [Diffuse desynchronized low voltage fast waves (under local anaesthesia) in cases of intractable pain].

(4) Endocrinological responses; Elevation of plasma level of pituitary hormones, catecholamines, nonesterified fatty acids; Elevation of CSF level of beta-endorphins.

Figure 2 shows typical responses elicited by high frequency stimulation. The area, where these sympathetic responses are obtained, actually forms a small triangle, in the lateral view, in the posterior hypothalamus. The triangle is delineated by lines connecting the mid-commissural point, the posterior commissure and the anterior border of the mammillary body (Fig. 3). Sano[10, 11] called it the "ergotropic triangle". In the anteroposterior view, it occupies a zone more than 1 mm and less than 5 mm lateral to the third ventricle wall.

Among these responses, elevation of blood pressure is the most reliable sign for confirming the proper position of the lectrode tip in the target point, because it is usually elicited in the most narrow area in the posterior hypothalamus as shown in Fig. 3[15]. If the tip of an electrode is inserted more medially or laterally outside the ergotropic triangle, stimulation usually causes decrease of blood pressure. Therefore, there seems to be a three-layer organization from medial to lateral in the posterior hypothalamus, namely the most medial parasympathetic zone, the middle sympathetic zone and the lateral parasympathetic zone[2, 15]. However, in the conscious state under local anaesthesia (in cases of intractable pain), decrease of blood pressure can rarely be observed. Instead, elevation of blood pressure is elicited by stimulation in the wide area of the posterior hypothalamus. In the conscious state high frequency stimulation causes a very unpleasant sensation. The patients report that a weak stimulation causes warm sensation in the whole body, but a strong one elicits a fearful feeling or horror with dizziness and rotating sensation. In cases of cancer pain, stimulation with weak parameters often produces pain relief. After confirming the proper placement of the electrode into the target point by means of x-ray and stimulation studies, electrocoagulation of the target is performed using the same bipolar electrode or a thermocontrol electrode (with a diameter of 2 mm, exposing its tip by 2 mm). High frequency current of 1 MHz, 2–4 watts is given for 30 seconds at 70 °C of the tip temperature. The coagulation lesions is about 3–4 mm in diameter, which was confirmed in the autopsied cases with cancer pain.

Table 1. *Follow-up Assessment of the Cases Undergoing Posteromedial Hypothalamotomy for Behavioural Disorders*

Effect on behavioural disturbances / State of living	Staying at home			Centres for the handicapped		Psychiatric institution	Total
	Part-time job	Household helping	Selfcare only	Vocational training	Unable to work		
Excellent	4	9	1	4	0	0	18 (49%)
Good	2	5	0	1	1	2	11 (30%)
Fair	0	1	0	0	3	1	5 (14%)
Poor	0	0	1	0	0	2	3 (8%)
Total	6	15	2	5	4	5	37 (100%)
	23 (62%)			9 (24%)		5 (14%)	

The operation on the other side is carried out, if necessary, in the same fashion after 2–3 weeks. Usually no neurological deficits are observed postoperatively except a temporary somnolent state.

Results

There were 60 cases of behaviour problems operated upon from 1962 to 1977. Most of the patients had a history of epileptic seizures associated with some degree of mental retardation, prior to the development of behavioural problems. This fact implies that the behavioural disturbances in these patients were also the results of cerebral dysfunction based upon some organic lesions. Pneumoencephalography often showed diffuse ventricular enlargement; porencephaly or any focal abnormality of the frontal lobe was not rarely seen. The types of behavioural disorders may be classified into three categories:

(1) Aggressive behaviour (usually against a particular person of the family or community, sometimes in the form of self-mutilation)

(2) Rage attacks (breaking objects, attacking anybody in the surroundings)

(3) Restless behaviour (constantly moving, running, jumping, leaping etc. sometimes referred to as hyperkinetic children)

For any type of these behavioural disorders, indication of this procedure was considered only when the behavioural problem became so severe that no medication was effective and isolation of the patient was necessary to prevent possible danger to the other members of the family or community. Among 60 cases (males 44 and females 16), 29 cases (males 22 and females 7) were under 15 years of age at the time of operation. There were two cases of reoperation and one case of hospital death due to haemorrhage from a vascular malformation. The follow-up studies were made in 1969[15], 1977[7, 8], 1985 and 1987. The results were fairly consistent in these different studies.

During the postoperative periods of 10 to 25 years in the 1987 follow-up, 13 patients had died of various causes: suicide three cases, status epilepticus two cases, accidental drowning two cases, hepatitis, pneumonia, renal failure, lung cancer, haematological disease and unknown cause one case each. Since any cause of death does not seem directly related to the surgical intervention, this unusually high rate of death may reflect a relatively short life expectancy of those handicapped patients. The most recent assessment of the effect of the operation upon behavioural disorders and of the present state of living of the patients is summarized in Table 1. In these 37 cases, precise information of their daily life was available by interviewing not only the members of their families but also professional people who had been taking care of these patients for a long time. There were excellent results in 18 cases (49%), no violent and aggressive behaviour postoperatively, and the patients had become so cooperative that familial and social adaptation had been established.

Eleven cases (30%) showed good results: no violent and aggressive behaviour, but still easily excited. As a whole, the results in 29 cases (78%) were considered to be satisfactory. There were five other cases with fair results: still occasionally showing aggressive behaviour, and three cases with poor results: apparently no clinical improvement. State of living of the patients was variable, mainly depending upon both postoperative behavioural problems and the degree of pre-existent mental retardation or epileptic seizures. There were 23 cases (62%) living at home. It was really encouraging to see that six patients had been employed, although doing part-time jobs and their income was not sufficient enough to live independently. Nine cases (24%) were staying at centres for the physically handicapped. However, five of them were able to work at the occupational training section in the centres. Five patients (14%) had still remained in the psychiatric institutions, due to poor social adaptation and severe mental retardation. The sedative effects of the operation can be estimated, once established, to be stable and consistent for as long as ten years or more. The importance of special education and care for the long postoperative period of time should be emphasized to restore their familial and social adaptation.

There have been reports of other series of relatively small numbers of patients[1, 4, 9, 18, 19, 20]. The results are similar to our study, with satisfactory clinical improvement in 80% or more of the patients. Memory disturbance and decreased spontaneity were reported in a few cases, as complications of this procedure. Intelligence quotient scores usually improved postoperatively, because the patients became able to concentrate on the tasks.

Endocrinological Check-up

Because the location of the surgical lesion in this procedure is close to the neuroendocrinological centres of the hypothalamic-pituitary system, the effects on the plasma values of various pituitary hormones were carefully studied. In conclusion, the stimulation of the posterior hypothalamus caused a temporary elevation of plasma values of many pituitary hormones, especially prolactin, lutenizing hormone and growth hormone; no significant changes of plasma values were noted postoperatively[7, 8]. Although several patients showed relatively low level of urine 11-OHCS and 17-KS, substitution therapy was not indicated. The postoperative physical development curve could be made from the data of ten patients, who were operated upon in their childhood. Although a few patients showed a tendency to gain weight after the operation, no serious developmental problem was observed (Fig. 4). The appearance of secondary sexual characteristics was normal. Menarche of the girls submitted to this operation was between 10 and 13 years of age. In seven cases, the follow-up stimulation tests of pituitary hormone release were performed, using the insulin test for growth hormone and ACTH, the LHRH test for LH and FSH, the TRH test for TSH and prolactin. The results were within normal range, to show that the hormonal regulation was not impaired even after the bilateral posteromedial hypothalamotomy. Before the operation, the plasma level of nonesterified fatty acids (NEFA) in these patients was relatively high in comparison with normal subjects. The stimulation of the posterior hypothalamus caused a temporary elevation of NEFA levels in the plasma[15]. Postoperatively the fasting level of NEFA became normal. These changes of NEFA levels before, during and after the operation may probably be related to the sympathetic responses of the patients, who usually showed a state of sympathicotonia in the pre-operative epinephrine test or mecholyl test, which returned to normal after the operation.

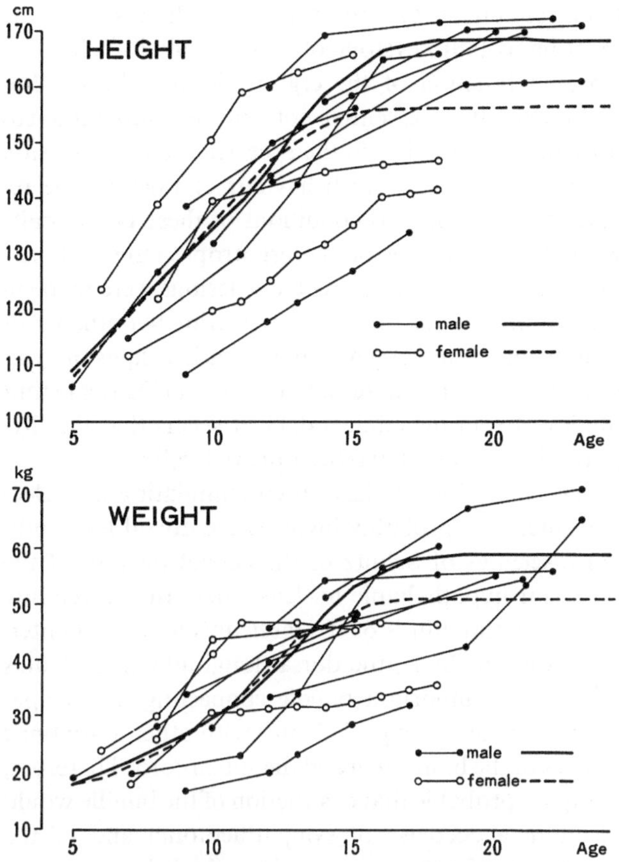

Fig. 4. Physical development curves of the childhood patients following posteromedial hypothalamotomy. Value at the youngest age of each curve indicates value at the time of the operation. Solid (male) and broken (female) lines mean the standard development curves for Japanese children

Discussion

According to Hess[5, 6], the diencephalon can be classified into two functional sectors, namely the ergotropic sector and the trophotropic sector. Connected with these two sectors, there seem to be two functional circuits in the prosencephalon, the ergotropic and the trophotropic circuits or systems, the hypothalamus being the most important keystone of these circuits[10, 11]. The integrated function of the ergotropic and the trophotropic circuits is essential not only in autonomic-endocrine activities, but in the experience and expression of emotion. Aggressive behaviour or rage can be regarded as unbalanced conditions of these two circuits with the dominance of the ergotropic circuit. Based upon this hypothesis, stereotactic lesions were made in the ergotropic portion of the posterior hypothalamus (ergotropic triangle) where most marked signs of sympathetic discharge were obtained upon electrical stimulation. From the clinical data, it seems that this hypothesis may well have been proved valid.

The area that we have been stimulating and electrocauterizing probably involves the dorsal longitudinal fasciculus of Schütz or the caudal portion of the posterior hypothalamic nucleus where the descending and ascending fibres of the fasciculus leave and enter. According to Ban[2], the dorsal longitudinal fasciculus is the most important bundle connecting the sympathetic zone of the hypothalamus with other autonomic centres of the brain stem and the spinal cord. Therefore, it is quite probable that destruction of the bundle would result in a decrease of sympathicotonia and of the expression of rage or aggression which is always accompanied by signs of sympathetic discharge. According to Tokizane *et al.*[21], the neocortex is activated by the reticular activating system with or without relay in the nonspecific thalamic nuclei, while the limbic system is activated mainly by the hypothalamic activating system, namely, the archicortex being activated by the posterior and the paleocortex by the anterior hypothalamus. The posterior hypothalamus also influences the neocortex through the reticular system. The latter is driven chiefly by impulses mediated in the somatic sensory system, and the hypothalamic activating system is driven by impulses in the visceral afferents and by humoral factors, such as epinephrine, norepinephrine, etc. Accordingly posteromedial hypothalamotomy will result in decreased activation of the limbic system, especially of the archicortex, and of a part of the neocortex. This will exert a profound influence upon emotion and behaviour mechanisms.

Although the desirability of such a procedure as this posteromedial hypothalamotomy in childhood is still controversial, the experiences of the present group with 29 children seem to support an early surgical intervention. There have been no serious developmental or endocrine side-effects in these cases, as long as the lesion did not involve the anterior hypothalamus. The obvious advantage of an early intervention is the possibility to eliminate the inevitable isolation of these children and to make it possible for them to get special, intensive education in schools or centres for the handicapped.

References

1. Balasubramaniam V, Kanaka TS (1975) Amygdalotomy and hypothalotomy—a comparative study. Confin Neurol 37: 195–201
2. Ban T (1966) The septo-preoptico-hypothalamic system and its autonomic function. In: Tokizane T, Schádе JP (eds) Prog Brain Res 21 A. Elsevier, Amsterdam, pp 1–43
3. Beattie J (1932) Hypothalamic mechanisms. Canad Med Ass J 26: 400–405
4. Black P, Uematsu S, Walker AE (1975) Stereotaxic hypothalamotomy for control of violent aggressive behaviour. Confin Neurol 37: 187–188
5. Hess WR (1947) Vegetative Funktionen und Zwischenhirn. Helv Physiol Pharmac Acta, Suppl IV
6. Hess WR (1949) Das Zwischenhirn. Syndrome, Lokalisationen, Funktionen. Schwabe, Basel
7. Mayanagi Y, Hori T, Sano K (1978) The posteromedial hypothalamus and pain-behaviour, with special reference to endocrinological findings. Appl Neurophysiol 41: 223–231
8. Mayanagi Y, Sano K (1979) Long-term follow-up results of the posteromedial hypothalamotomy. In: Hitchcock ER, Ballantine HT Jr, Meyerson BA (eds) Modern concepts in psychiatric surgery. Elsevier North-Holland Biomedical Press, pp 197–204
9. Rubio E, Arjona A, Rodriguez-Burgos F (1977) Stereotactic cryohypothalamotomy in aggressive behaviour. In: Sweet WH *et al* (eds) Neurosurgical treatment in psychiatry, pain and epilepsy. University Park Press, Baltimore, pp 439–444
10. Sano K (1962) Sedative neurosurgery with special reference to posteromedial hypothalamotomy. Neurol Med Chir (Tokyo) 4: 112–142
11. Sano K (1966) Sedative stereoencephalotomy: fornicotomy, upper mesencephalic reticulotomy and postero-medial hypothalamotomy. In: Tokizane T, Schádе JP (eds) Prog Brain Res 21 B. Elsevier, Amsterdam, pp 350–372
12. Sano K (1974) Surgery of the hypothalamus—in commemoration of Otfrid Foerster. In: Sano K, Ishii S (eds) Recent progress in neurological surgery. Excerpta Medica, Amsterdam, pp 210–218
13. Sano K (1975) Posterior hypothalamic lesions in the treatment of violent behaviour. In: Fields WS, Sweet WH (eds) Neural bases of violence and aggression. WH Green, St. Louis, pp 401–420
14. Sano K (1982) Aggressiveness. In: Schaltenbrand G, Walker AE (eds) Stereotaxy of the human brain. Thieme-Stratton, New York, pp 617–621
15. Sano K, Mayanagi Y, Sekino H, Ogashiwa M, Ishijima B (1970) Results of stimulation and destruction of the posterior hypothalamus in man. J Neurosurg 33: 689–707
16. Sano K, Sekino H, Hashimoto T, Amano K, Sugiyama H (1975) Posteromedial hypothalamotomy in the treatment of intractable pain. Confin Neurol 37: 285–290
17. Sano K, Sekino H, Mayanagi Y (1972) Results of stimulation and destruction of the posterior hypothalamus in cases with

violent, aggressive or restless behaviours. In: Hitchcock E *et al* (eds) Psychosurgery. Thomas, Springfield, Ill, pp 57–75

18. Schvarcz JR (1977) Results of stimulation and destruction of the posterior hypothalamus: a long-term evaluation. In: Sweet WH *et al* (eds) Neurosurgical treatment in psychiatry, pain and epilepsy. University Park Press, Baltimore, pp 429–438
19. Schvarcz JR, Driollet R, Rios E, Betti O (1972) Stereotactic hypothalamotomy for behaviour disorders. J Neurol Neurosurg Psychiatry 35: 356–359
20. Sranja M, Nádvornik P (1975) Surgical complication of posterior hypothalamotomy. Confin Neurol 37: 193–194
21. Tokizane T, Kawamura H, Imamura G (1960) Hypothalamic activation upon electrical activities of paleo- and archicortex. Neurol Med Chir (Tokyo) 2: 63–76

Correspondence: K. Sano, M.D., Prof. Emeritus of Neurosurgery, Departments of Neurosurgery, Teikyo University and University of Tokyo, Japan.

Acta Neurochirurgica, Suppl. 44, 152–157 (1988)

Stereotactic Operation in Behaviour Disorders. Amygdalotomy and Hypothalamotomy

B. Ramamurthi

Department of Neurosurgery, Dr. A. Lakshmipathi Neurosurgical Centre, VHS Medical Centre, Madras, India

Summary

After a survey of the anatomical and physiological basis of operative treatment of behaviour disorders by stereotactic lesions in the amygdala and the posterior medial hypothalamus the author describes his own experiences with 603 operations for control of conservatively untreatable aggressiveness. In 481 cases bilateral amygdalotomies and in 122 mostly secondary posteromedian hypothalamotomies have been performed. Initially excellent or moderate improvement was achieved in 76%. After a follow-up of more than three years this figure only slightly decreased to 70%. The group of patients who did not positively respond (30%) needs further study to discover the reasons for failure.

Introduction

Our understanding that behaviour disturbances can be caused by disorders of the brain is barely two centuries old. Mental and behavioural disorders were attributed to disturbances in vapours or humours in the body or to supernatural influences. Amulets and Charms were more in demand than drugs and physicians who anyway were ineffective.

Macbeth: Canst thou not minister to a mind diseas'd; Pluck from the memory a rooted sorrow; Raze out the written troubles of the brain; And with some sweet oblivious antidote, cleanse the stuff'd bosom of that perilous stuff, which weighs upon the heart?
Doctor: Therein the patient must minister to himself.
Macbeth: Throw physic to the dogs, – I'll none of it.

Shakespeare V. iii. 40–48

Gradual realization of the association of behaviour patterns with neural structure was achieved by careful clinical and experimental observations during the 19th and early 20th century, and the latter half of the 20th century has witnessed tremendous advances in neurosciences.

In this milieu, it is essential for the neurosurgeon to be on solid grounds about the physiological basis on which his operations are performed – especially with regard to functional neurosurgery aimed at modifying behaviour. Such operations have come in for severe, unjust and uninformed criticism in the western countries and arguments are offered against making lesions in the so-called normal areas of the brain to help behaviour disturbances. Peculiarly a similar outcry has not been raised against operations for epilepsy where also lesions may be made in "normal areas of the brain". This was referred to as psychological diplopia by Mindus, one of the speakers in the symposium on personality arranged by the Eurasian Academy of Neurosurgery at Brussels in August 1987. Clearer elucidation of the neurobiological basis of functional neurosurgery is an important step in dispelling prejudices against a useful type of neurosurgery. Surgical procedures on the brain to control restless or aggressive behaviour have been included under the term sedative neurosurgery (Sano 1966).

Early attempts at controlling behaviour disorders by ablation of the frontal or temporal lobe were not uniformly successful and also led to undesirable results. This was inevitable as knowledge regarding the anatomical and physiological basis of behaviour was meagre. The understanding of the various behavioural mechanisms was hampered by the fact that the part of the brain which, as we know today controls behaviour was labelled the rhinencephalon or olfactory brain and was considered vestigeal in man. The breakthrough came when Bard (1928) and Cannon (1929) following Woodworth and Sherrington (1904) established that the "sham rage" phenomenon arose from the region of the hypothalamus. Papez in 1937 proposed the concept of a circuit controlling emotions in the medical portion of the cerebral hemisphere. By correlation of the clinical and laboratory data, he suggested that "the hypothalamus, the anterior thalamic nuclei, the gyrus cinguli, the hippocampus and their interconnections constitute a harmonious mechanism which may elaborate the functions of central emotion, as well as participate in emotional expression".

The amygdala was not included in this concept of the behavioural brain. During the past two decades, however, it has become increasingly obvious that the amygdala plays a pivotal and vital role in the production of behaviour patterns through its connections to the cortex, the limbic system, the hypothalamus and thalamus.

Whereas the circuit that Papez proposed is essentially on the medial side of the hemisphere aligned parallel to the sagittal plane, Livingston (1969) proposed another limbic circuit at right angles to this, termed the basolateral circuit and consisting of the "Orbito-insular temporal connections". The medial circuit of Papez connects to the anterior thalamic nucleus while the basolateral circuit connects to the dorsomedian nucleus. The basolateral circuit reaches the thalamus without the interposition of a hypothalamic midbrain relay. Livingston (1969) suggested that normal behaviour may be the result of correct balance between the basolateral and the medial circuits.

The behavioural brain may be conceived as having a central core of executive structures with extension and distant connections. The amygdala, the hypothalamus and the periaqueductal grey form the central core; the inferior orbital cortex, the cingulum, the dorsomedian nucleus of the thalamus and the internal medullary lamina may be considered as extensions of this system. The inferior orbital cortex is connected with the amygdala by the uncinate fasciculus and functions as a frontal extension of this system (Nauta 1964). The cingulum is connected to the limbic nuclei of the thalamus and is more concerned with emotional "feeling" (a purely subjective phenomenon) than with emotional behaviour. The dorsomedian nucleus of the thalamus by virtue of its connections with the cingulum and hypothalamus forms an integral part of the behavioural brain on the "affect" side. The anterior third of the internal medullary lamina connects to the hypothalamus, and is involved in the modification of behaviour.

The close proximity of the olfactory brain and the behavioural brain and their interconnections was not only an anatomical accident but represents an important stage in functional evolution to man from lower animals in whom olfactory sensations were major determinants of behaviour. Thus, in man also, the olfactory and non-olfactory parts are closely interrelated at many levels including the amygdala.

The Amygdala

The amygdaloid nucleus has a dural origin. It develops partly from the primitive olfacto-striatum and partly from the hypopallium. It has been identified in cyclostomes and tailless amphibians. It reaches an unsuspected complexity in man and as we go up the evolutionary scale, it undergoes an increase in size, until it reaches a maximum size of 1,200 mm^3 in man. The terminology used to describe the amygdaloid nucleus is a little confused because the same descriptive terms used for lower animals have been applied to man, without taking into consideration the changes in the position of the amygdala. The amygdaloid primordium lies on the floor of the temporal horn. As the temporal horn changes in form, the amygdala is carried medially and also rotates till it eventually (in man) lies on the antero-superior wall exactly opposite the inferolateral bulge of the hippocampus. It is this shift that causes some confusion when terms like corticomedial and basolateral are used.

The amygdaloid nucleus lies in close approximation to the superomedial wall of the temporal horn. In sagittal sections the centre of the amygdaloid nucleus is about 4.5 mm in front of the anterior most point of the temporal horn; its long axis lies along the roof of the temporal horn. In the coronal plane, it is related to the medial margin of the temporal horn.

The Structure of the Amygdala

Many types of subdivisions of the amygdala are in vogue. Broadly speaking the amygdala can be divided into the basolateral (which is non-olfactory) and corticomedial (which is olfactory). The main subdivisions have different afferent and efferent connections. The connections of the amygdala are to the temporal lobe (amygdalo-polar bundle), the hypothalamus (the stria terminalis and the ventral amygdalo-hypothalamic tract), the thalamus and the frontal lobe. All these connections though predominantly one-way, are two-way circuits. The inferior orbital cortex of the frontal lobe is considered to be a frontal extension of the limbic system. The uncinate fasciculus arises from the temporal pole and partly from the amygdala and spreads into the inferior orbital cortex.

A few words about the anatomy of the hypothalamus would be relevant here. Two important fibre tracts course through the hypothalamus, viz., the medial forebrain bundle connecting the various areas of the limbic system and the mesencephalon, and the fornix, an efferent pathway from the hippocampus to the mamillary body, consisting of many fibres to the hypothalamic nuclei. The afferent connections of the hypothalamus are: 1. the amygdalo-hypothalamic pathways that have already been described, 2. fronto-hypothalamic fibres and 3. fibres from the ventral anterior nucleus of the thalamus to the hypothalamus. For purpose of behavioural physiology, the hypothalamus can be divided into two zones – the medial-ergotrophic and the lateral trophotrophic zones, the medial ergotrophic corresponding to the dynamogenic and the lateral to the adynamogenic zones of Hess (1954). Most of the amygdaloid connections end in the medially situated ergotrophic zone. Stimulation of this zone (particularly near the perifornical nucleus) causes the appearances of sham rage.

Basis for Operation

Theoretically a surgical lesion anywhere in this system should help to modify behaviour disorders. However, only lesions in certain areas afford maximum relief from behavioural alterations. The two areas usually selected for operation are the amygdaloid nucleus and the hypothalamus.

Aetiology of Behaviour Disturbances

A disturbance of the delicate balance between various neuronal circuits in the hypothalamo-limbic system results in abnormal behaviour patterns. While such disturbances may occur temporarily from injury, drugs, toxins etc, more lasting trouble results from epileptic disorders and encephalitis.

Post-epileptic behavioural disorders may occur in a patient who has been a long standing sufferer from grand mal seizure. The behavioural disorder often sets in some time after adequate control of the seizures by anti-convulsant medication.

Post-encephalitic behavioural disorders may follow immediately or some months after an attack of encephalitis. The encephalitis can only be surmised from a history of high fever and altered sensorium with or without generalized seizures. The other aetiological factors include trauma, meningitis, vascular disorders and schizophrenia. In these cases the behaviour disorders occur without epilepsy.

Types of Behaviour Disorders

1. Hyperkinesis, restlessness and wandering tencency.

2. Destructive and violent tendencies – These may be continuous or intermittent and may arise with or without provocation.

3. Pyromania – A few patients have an obsessive desire to set fire to things, inanimate or animate.

4. Self-destruction: Some of the patients exhibit a marked tendency to hurt themselves.

Stereotactic amygdalotomy was first introduced by Narabayashi *et al.* in 1963 for control of abnormal behaviour such as aggressiveness, violence, restlessness etc. Narabayashi suspected that the selective and severe changes seen in the hippocampus in epileptic brains may induce functional disturbance in the neighbouring amygdala, resulting in emotional irritability and "epileptic" personality changes. He was also led by experiments on this nucleus in animals to perform operation on the amygdala (Ward 1948, Kaada 1952, Maclean and Delgado 1953).

Stereotactic hypothalamotomy was first done by Spiegel and Wycis in a schizophrenic patient (1962).

Keiji Sano pioneered the operation of hypothalamotomy for behaviour disorders and reported a series of 22 patients in 1966 and 51 cases in 1970.

For many years we in Madras (Balasubramaniam and Ramamurthi 1970) following the lead given by Narabayashi and Sano have been interested in a study of the role of the amygdala and the hypothalamus in the modification of behaviour and have made lesions in the amygdala and hypothalamus to control varieties of aggressive behaviour disorder.

Unilateral or Bilateral Operations

Unilateral amygdalotomy or hypothalamotomy may be done if the intensity of behavioural disturbances is moderate, but most cases require bilateral amygdalotomy or bilateral hypothalamotomy. Operations for behavioural (or emotional) disorders have generally to be done on both sides since there is no laterality of emotional feeling or emotional behaviour. One condition in which a strictly unilateral procedure (either amygdalotomy or hypothalamotomy) gives good relief is the behaviour disorder associated with infantile hemiplegia. These patients who were once candidates for hemispherectomy are considerably benefited by unilateral sedative neurosurgery (Balasubramaniam 1980).

One Stage or Two Stages

Bilateral amygdalotomy may be done in two successive stages. But it has been shown that bilateral, one-stage amygdalotomy can be done safely and is not attended with extra defects attributable solely to the bilaterality of the procedure. Bilateral hypothalamotomy is done only in successive stages (Sano 1970, Balasubramaniam *et al.* 1973).

Amygdala First or Hypothalamus First

Some surgeons prefer hypothalamotomy as the procedure of choice (Sano 1970), while others prefer amygdalotomy as the first step (Narabayashi 1963, Balasubramaniam *et al.* 1967, Heimburger 1966) for the following reasons (Balasubramaniam *et al.* 1973): 1. the amygdala is considered to be the "higher" centre, 2. bilateral amygdalotomy is a one-stage operation whereas hypothalamotomy has to be done in two stages and 3. elimination of the amygdala is perhaps more logical as in many cases of behaviour disorders, there is likely to be a defect in the amygdala; full proof for this is however lacking.

In patients in whom bilateral amygdalotomy does not succeed, hypothalomotomy is performed. Hypothalamotomy done after amygdalotomy has been termed secondary hypothalamotomy and in such cases unilateral lesions in the hypothalamus are usually effective (Balasubramaniam 1973).

Personal Material

Out of a total of 1,774 stereotactic operations performed over a period of 28 years, 603 operations have been done for control of aggressive behaviour disorder. Of these 481 were bilateral amygdalotomies and 122 were posteromedian hypothalamotomies.

Most of the patients were children below the age of 15 who had developed aggressive behaviour disorder of restlessness as a result of some insult to the brain. The types of behaviour problems included physical aggression, hyperkinesis, wandering tendency, destructive and self-destructive tendencies. In some instances, the behaviour disorder was associated with epilepsy or developed as a sequel to epilepsy, while in others there was no such history or association with epilepsy. All these children had received medical treatment with modern psychotropic drugs over a period of two years without any appreciable relief.

In the initial stages, patients with reasonable intelligence were chosen; later, even if the IQ was low, the patients were offered operation to cure their violence or restlessness so that their parents and siblings might be relieved of constant anxiety and unhappiness. In the early stages, only older children were chosen for operation, but with the experience gained, we now offer this form of treatment even to children who are six or seven years old, provided a minimum of two years of intensive medical treatment has been unsuccessful. No purpose is served by postponing operation and the earlier the child is rid of the behaviour problem, the greater are the opportunities of possible future education.

Preoperative and postoperative psychological assessment was possible in only 60 of these children. It was difficult to quantify the restlessness. Narabayashi uses a room where the floor is marked into different blocks; the number of blocks the child traverses over a fixed time period indicates the intensity of the restlessness.

EEG studies did not provide any useful criteria for the decision regarding operation or to determine prognosis, except that when there was evidence of grand mal or temporal lobe epilepsy, the results were better. The first choice of operation was bilateral simultaneous stereotactic amygdalotomy (402 cases). In those cases where this did not succeed, hypothalamotomy was performed (secondary hypothalamotomy 73 cases). This was usually 8–12 weeks later. In 47 cases, hypothalamotomy was performed as the primary operation.

Surgical Procedure

The operations were done under general anaesthesia using the Leksell stereotactic frame. The approach was through coronal burr holes. The coordinates for the amygdala were calculated after demonstrating the tip of the temporal horn by myodil ventriculography.

After anatomical localization, verification of the target was obtained by the effects of stimulation. The ideal area in the amygdala seems to be one that gives the maximum response of apnoea. Depth recordings from the amygdala were useful only in a small percentage of cases. The target area in the posteromedian hypothalamus was identified by stimulation effects such as rise of blood pressure, pulserate, apnoea and ocular movements (Balasubramaniam *et al.* 1973, Kalyanaraman 1975).

The amygdalar lesions were made with a combination of diathermy and Myodil wax and were about 600–900 mm^3 in size. The hypothalamic lesions were made with heat or radio frequency waves and were about 120–200 mm^3 in size.

In 6% of cases of amygdalotomy there was an immediate or slightly delayed hemiplegia which recovered over a few weeks. This may be related to the route of introduction of the electrode through coronal burr holes.

Results

After bilateral amygdalotomy 39% of patients showed good to excellent and 37% moderate improvement in restlessness and behaviour disorder. A similar percentage of patients were benefited by primary hypothalamotomy. In those patients in whom amygdalotomy failed and was followed by "secondary" hypothalamotomy, half showed improvement.

The benefits obtained from operation were assessed from three points of view.

1. Improvement in restlessness or violence: The results was considered good or excellent when the patient continued to be calm and quiet, despite provocation. Moderate improvement implies a diminution or absence of aggression or restlessness when there was no provocation.

2. Beneficial effect of a quiet patient on the patients' siblings and relatives: The immeasurable value for the family is indicated by the response from parents and relatives, whose quality of life suddenly improved, and also from the increasing demand for such operations.

3. Better possibilities of educating the child result from the improvement in restlessness. Preoperative psychological tests are difficult because of the restlessness and inability to concentrate. Postoperatively, repeated psychological testing was possible in many different children during long term follow-up.

A long term follow-up of more than three years showed the final pattern of the benefits obtained (Ramamurthi 1977). Of all the cases operated on 55% continued to maintain good improvement and 15% moderate improvement, whereas 30% of patients failed to respond to either amygdalotomy or hypothalamotomy or to both. Heimburger (1978) found that 50% of patients operated on primarily for seizures, 33% for uncontrolled conduct disorders and 50% with both conditions seemed improved after operation. *It is the failed group which needs further study to discover why they did not respond to operation.*

A study of gastric acid secretion indicates that gastric acid levels may prove to be a factor of prognostic significance (Ramamurthi 1975). In those cases in whom the gastric acid levels were high initially, but were lowered after a stereotactic operation, the postoperative effect on the behaviour disorders appeared better. Another approach has been suggested by Heimburger (1971, personal communication). He found that an absence of variation in the diuranal cortisol level may indicate poor results after operation. These areas need further investigation.

Conclusions

Our results show that a stereotactic operation benefits some 60% of children with aggressive behaviour disorders. When such disorder is associated with epilepsy, the chances of improvement are higher than when this is not the case. There is no deterioration in the behaviour or the IQ. About a third of the cases do not benefit and this group needs further study. This form of operation has not been used by us on adult patients with aggressive tendencies. However, such patients may also benefit from a stereotactic operation if this is associated with epilepsy.

Until such time as the chemistry of behaviour is better understood and drug treatment becomes fully effective, stereotactic operations in the control of aggressive behaviour disorder will continue to have a useful place.

From the above, it is obvious that lesions in the amygdala do modify behaviour towards the better either alone or in combination with a lesion in the hypothalamus. However the areas of our future interest should be in those cases whom we could not help. What are the factors that prevented a good outcome in some of these cases? Such a study will add to our knowledge of the basic mechanisms.

With this experience, it is strange to note that in the Western neurosurgical world, these operations have not become popular.

In conclusion one must pose an important question to this elite collection of intelligentsia who have gathered here to discuss human behavioural aspects. It is a well known fact that no animal kills its own kind. Even in a battle for supremacy of the herd, the vanquished runs away and the victor does not puruse the vanquished and destroy it. Our brains have evolved for the animal kingdom over a million years. Where from did aggression against its own kind come into human behaviour. How did such a mutation occur and when? Why this inbuilt driving force that makes humans hate each other for flimsy reasons and make all our efforts to destroy them? This is a question that anthropologists, sociologists and neurologists will soon have to answer, if there is to be any hope for the future of the human race.

References

1. Balasubramaniam V, Ramamurthi B, Jagannathan K, Kalyanaraman S (1967) Stereotaxic amygdalotomy. Neurol India 15: 119–122
2. Balasubramaniam V (1980) Functional and sedative neurosurgery. In: Ramamurthi B, Tandon PN (eds) Text book of neurosurgery. Grient Longmanns Ltd, Delhi, Madras, pp 1112–1138
3. Balasubramaniam V, Kanaka TS, Ramanujam PB, Ramamurthi B (1973) Stereotaxic-hypothalamotomy. Confin Neurol (Basel) 35: 138–143
4. Bard P (1928) A diencephalic mechanisms for expression of rage with special reference to sympathetic nervous system. Am J Physiol 84: 490–515
5. Cannon WB (1963) Bodily changes in pain, hunger, fear and rage. An account of recent researches into the function of emotional excitment, 2nd ed. New York
6. Heimburger RF, Small IF, Small JG, Milstein V, Moore D (1978) Stereotactic amygdalotomy for convulsive and behavioural disorders. Applied Neurophysiology 41: 43–51
7. Hess WR (1954) Diencephalon, autonomic and extra-pyramidal functions. Grune and Stratton, New York
8. Kaada BR (1951/1952) Somato-motor, autonomic and electrocorticographic responses to electrical stimulation of "rhinencephalic" and other forebrain structures in primates, cat and dog. Acta Physiol Scand [Suppl] 24

9. Kalyanaraman S (1975) Some observations during stimulation of the human hypothalamus. Confin Neurol (Basel) 37: 189–192
10. Livingston K (1972) The frontal lobes revisited. The case for a second look. Arch Neurol 20: 90–95
11. Maclean PD, Delgado JMR (1953) Electrical and chemical stimulation of fronto-temporal portion of limbic system in the waking animal. Electroenceph Clin Neurophysiol 5: 91–100
12. Narabayashi H, Nagao T, Saito Y, Yoshida M, Nagahata M (1963) Stereotaxic amygdalotomy for behaviour disorders. Arch Neurol Chicago 9: 11–26
13. Narabayashi N (1973) Stereotaxic operations for behaviour disorders. Progress in neurological surgery 5: 113–158
14. Nauta WJH (1964) Some efferent connections of the prefrontal cortex in the monkeys. In: Warrent JM, Akert K (eds) Frontal granular cortex and behaviour. McGraw Hill Book, New York
15. Papez JW (1937) A proposed mechanism of emotions. Arch Neurol Psychiatry 38: 725–743
16. Ramamurthi B, Mascreen M, Valmikinathan K (1977) Role of the amygdala and hypothalamus in the control of gastric secretion in human beings. Acta Neurochir (Wien) 24: 187–190
17. Ramamurthi B (1977) Surgery for aggressive behaviour disorders. International Congress Series No. 433. Proceedings of the Sixth International Congress of Neurological Surgery, Brazil, pp 19–25
18. Sano K (1966) Sedative stereoencephalotomy; fornicotomy, upper mesencephalic reticulotomy and posterior medical hypothalamotomy. In: Tokiazene T, Schade JP (eds) Progress in brain research 21-B: Correlative neurosciences. Part B. Clinical studies. Elsevier Publishing Company, Amsterdam New York, p 350
19. Sano K, Mayanagi Y, Sekino H, Ogashiwam M, Ishijima (1970) Results of stimulation and destruction of the posterior hypothalamus in man. J Neurosurg 33: 689–707
20. Spiegel EA, Wycis HT (1962) Stereoencephalotomy. Part II, Clinical and physiological application. Grune and Stratton, New York
21. Ward A Jr (1948) The cingular gyrus; Area 24. J Neurophysiol 11: 13–23
22. Woodworth RS, Sherrington C (1904) A pseudoaffective reflex and its spinal pathology. J Physiol (London) 31: 234–243

Correspondence: Prof. B. Ramamurthi, M.D., Head of the Department of Neurosurgery, Dr. A. Lakshmipathi Neurosurgical Centre, VHS Medical Centre, Madras 113, India.

Acta Neurochirurgica, Suppl. 44, 158–162 (1988)

Psychosurgery Today

L. V. Laitinen

Summary

A review on indications, target points and results of stereotactic operations for treatment of psychiatric diseases is given, based on personal experiences and reports in the literature. As a conclusion the author suggests that the anatomical target should be chosen selectively. There is strong evidence that different approaches lead to different results. Cingulotomy is effective for chronic pain with addiction and depression, anterior capsulotomy for obsessive-compulsive and anxiety neurosis, innominotomy for chronic and recurrent depression, and postero-medial hypothalamotomy for restless, aggressive and destructive behaviour. Therefore, the target should be selected according to the individual symptoms of the patient. The results of operation are usually good and most patients can return to a normal life. The side-effects are infrequent and seldom serious. Modern psychosurgery does not modify the personality of the patient. On the contrary it often relieves it from disturbing symptoms of illness.

Keywords: Stereotactic surgery; psychosurgery; capsulotomy; cingulotomy; hypothalamotomy.

Introduction

Human stereotactic surgery began to develop in the 1940's when it had become evident that the beneficial effects of frontal lobotomy were obtained with lesions in the medial parts of the frontal lobes, such as the cingulum and the anterior limb of the internal capsule. John Fulton[4] of Yale University was probably the first who proposed that neurosurgeons should abandon the classical lobotomy and its modifications. They should instead place well localized and restricted lesions in the cingulum. This may have been an important impetus for the development of human stereotactic surgery which was introduced in 1947 by Spiegel *et al.*[14]. Soon after the new technique had become available, a rapid development of neuroleptica seemed to relegate psychosurgery to the past. In the late 1950's it was only practised in very few places. Leksell[10] in Lund developed a stereotactic technique for making lesions in the anterior branch of the internal capsule. Knight[7] placed radioactive Yttrium^{90} rods in the orbitofrontal white matter and Foltz and White[3] made thermocoagulation lesions in the anterior cingulum. Sano[13] began to treat aggressive and restless oligophrenic patients with lesions in the postero-medial hypothalamus. Narabayashi *et al.*[12] in 1963 treated oligophrenic and epileptic patients with restless and aggressive behaviour in the amygdala.

Stereotactic psychosurgery was practised widely in the late 1960's and in the early 1970's. Several well documented clinical reports indicated that the results were superior to those of previous open lobotomy. The surgical morbidity was very low and the mortality was practically nil. Blunting of the mental functions was seldom observed; on the contrary many patients regained their ability to feel pleasure. In the 1970's, however, strong opposition began to rise in many countries against psychosurgery. Often, the criticism lacked scientific ground and only reflected social mistrust and fear for a possibility of manipulation of the mind, but it could also be justified. Some neurosurgeons had proposed that homosexual persons with sexual offenses should be given a chance of ventromedial hypothalamotomy as an alternative to prison. They postulated that hypothalamotomy would "cure" the homosexuality and replace it by a less dangerous heterosexuality. Other neurosurgeons had suggested that prisons could be emptied if criminals underwent brain surgery. These uncritical ideas evoked very strong reactions among intellectuals and psychosurgery became more or less forbidden in several countries, *e.g.*, Germany, USA and Japan. In other countries, *e.g.*, Finland, Sweden, England, Spain, India and Australia, psychosurgery was still practised. Its pros and cons were discussed openly. Gradually psychosurgery, even when used in a limited number of patients, became an established form of treatment for some intractable psychiatric disorders in which all conventional conservative treatments had failed.

In these countries there was a good cooperation between psychiatrists and neurosurgeons. Psychosurgery was not considered to be a surgical, but a psychiatric problem. Thus it was always the psychiatrist who asked the neurosurgeon whether surgery might help a patient with a intractable and hopeless illness. Such a relation between the psychiatrist and the neurosurgeon evidently created a good basis for the development of psychosurgery. In the early 1970's, some plans were made to conduct controlled trials on the effects of psychosurgery, but it turned out that the task was too difficult.

Selection of the Patients

Stereotactic psychosurgery may be indicated in severe psychiatric disorders where conventional psychiatric treatment, including psychotherapy, psychoanalysis, drug therapy and electronconvulsive therapy have failed. The decision to consider psychosurgery rests on the psychiatrist. Since psychosurgery may still be considered to be experimental and is controversial it should only be practised in close cooperation with good psychiatrists and psychologists. I do not feel that laymen or priests have any competence to assist in the selection of patients. Patients with obsessive-compulsive neurosis, severe anxiety neurosis, depression, and intractable states of restlessness and agitation may be the best candidates for psychosurgery. It has been shown that psychotic patients may react as favourably as those with neurosis. Patients with chronic pain states combined with psychopathological features may also be candidates for operation.

Surgical Targets

Anterior Capsulotomy

In 1952 Lars Leksell began to make stereotactic lesions in the anterior limb of the internal capsule, just between the head of the caudate nucleus and the anterior putamen. Bilateral lesions were aimed at interrupting frontothalamic fibres. Between 1952 and 1957 Leksell operated on 117 patients who suffered from various types of mental illness. Leksell's original target lay ca. 20–25 mm lateral to the midline, about 17 mm rostral to the anterior commissure and extended from the level of the prolonged intercommissural line 20 mm in an oblique rostral and dorsolateral direction. In 1961, Herner[5] made a long-term follow-up study on these patients. He found that capsulotomy had been very effective in patients with schizophrenia and obsessive-compulsive neurosis; 85 and 78%, respectively, of the patients had shown a good or fair improvement. However, the degree of improvement was better in the obsessive than in the schizophrenic patients. Depressive patients had an outcome almost as positive: 74% had improved. The side-effects consisted of transient confusions and urinary incontinence, and a tiredness lasting for several weeks. Three patients out of the 117 showed serious long-lasting side-effects of personality changes resulting in conflicts with the law (overt sexual behaviour, alcohol addiction, and thefts, respectively).

Capsulotomy has since then become the most common psychosurgical procedure in Stockholm, from where Bingley *et al.*[2] have published several well documented clinical studies. Björn Meyerson *et al.*[11] were the first who began to place the capsular lesions with the aid of a CT study. They found that air ventriculography often caused ventricular dilatation, which made it difficult to place the lesions correctly in the narrow white matter between the head of the caudate nucleus and the putamen. I have used the CT as the only targeting method in psychosurgery since March 1984, and I feel that a CT or MRI study is necessary for an accurate placement of the lesions. The target lies usually 18–20 mm from the midline and 17 mm in front of the anterior commissure. The ventralmost part of the lesion lies at the height of the intercommissural line, from where it extends about 16 mm in frontal and dorsolateral direction, following the course of the internal capsule. The width of the lesions should be about 8 mm.

I perform almost all psychosurgery under local anaesthesia and without sedation, which permits well controlled electrical stimulation tests during the operation. The patient lies comfortably on the table and can usually perform properly during stimulation and give reliable answers. Nevertheless, fake stimulations are needed to make sure that the response reported is specific to stimulation.

High-frequency stimulation of the anterior capsule causes relatively often, in my own series in 23% of the patients, a decrease of anxiety and tension. Extrapyramidal responses, such as a sensation of trembling in the contralateral side of the body, are reported sometimes, but in the majority of the patients there are no subjective or objective reactions to stimulation.

The clinical results of capsulotomy are usually good for obsessive-compulsive neurosis. More than half of the patients are completely relieved of their obsessions. The anxiety, tension and depression also improve, but the grade of improvement is usually not so good as for obsessionals. About half of the patients can regain a normal working capacity. Crazy phobias may improve, but usually the result is not very good in phobic neurosis.

The side-effects of capsulotomy consist of a short-lasting confusion with occasional loss of bladder control lasting for a few days. There is usually a marked tiredness and a lack of initiative for several weeks. Vilkki[16] has found that even ten years after operration, capsulotomy patients have less initiative than those operated on in other psychosurgical targets.

Cingulotomy

In 1962, Foltz and White[3] carried out stereotactic lesions in the anterior cingulate bundle for chronic intractable pain. Thereafter a large series of cingulotomies have been carried out, most of them in the USA, and favourable results have been published[1]. Most surgeons place the lesions in the region of the middle anterior cingulum, about 2–4 cm behind the knee of the corpus callosum. The size of the bilateral lesions is about 10 mm in a dorsal and 12–14 mm in a lateral direction so that the whole fasciculus cinguli will be cut. Ballantine and coworkers[1] reported that extensive cingulotomy was effective for chronic pain combined with depression and addiction, and for anxiety neurosis, whereas the results were less good for obsessive-compulsive neurosis. In his patients two operations were often needed. The post-mortem studies in some of them showed that the lesions had not only interrupted the cingulate bundle, but they had also included parts of the corpus callosum and supracingulate frontal white matter. Kelly *et al.*[6] in London were evidently not happy with cingulotomy alone because they added bilateral lesions in the medio-basal frontal white matter. They called their operation "limbic leucotomy", which was reported to have been very effective for obsessive-compulsive neurosis.

Gunvor Kullberg[8], who compared the results of cingulotomy with those of anterior capsulotomy, found that cingulotomy, when it was clearly restricted to the fasciculus cinguli, was less effective for anxiety neurosis and much less effective for obsessive-compulsive neurosis than capsulotomy. Also my personal experience based on 55 patients operated on the cingulum is similar: cingulotomy is not effective for obsessive-compulsive neurosis.

Electrical stimulation of the cingulum often increases the anxiety and tension. Automatisms, in the form of unconscious mano-oral movements were seen in two of my six patients. Stimulation of the rostral cingulum, a common target in the beginning of my psychosurgical activity often caused an increase of anxiety and tension, but in four patients a decrease of these symptoms was obtained. I think that in these patients the stimulation response may have derived from the knee of the corpus callosum.

The US Congress review in 1977 of 200 cingulotomized patients showed that the approach had been very effective in patients with chronic pain combined with drug addiction and depression[15]. About 90% of such patients had become symptom-free. These cingulotomies had been carried out by Ballantine[1] and, as I already pointed out, the lesions may not have been restricted to the cingulum, but had presumably also included parts of the medial frontal white matter above the cingulum.

Cingulotomy is seldom accompanied by side-effects. Kullberg[8] in her comparative study noticed that slight confusion and affective deficits were clearly less frequent and also less severe than after anterior capsulotomy. Vilkki[16] in Helsinki has repeatedly shown that cingulotomy causes a decline in visual imagery. This finding gets support from that of Teuber *et al.*[15], who found that cingulotomized patients had problems in visual labyrinth tests. Since this study was retrospective and the authors had not studied the patients before surgery they could not be sure that the deficit was related to the cingulotomy.

Mesoloviotomy

Bilateral lesions in the knee of the corpus callosum (mesolovion in Greek) were performed by me in Finland[9]. The 6 by 8 mm large lesions in the rostral layer of the knee, 6 mm from the midline, were thought to interrupt interhemispherical cingulo-striate pathways. Electrical stimulation of the target with 60 Hz often caused a disappearance of anxiety and tension in schizophrenic patients. The symptoms gradually returned within a minute after cessation of stimulation. The response could be reproduced repeatedly. Only schizophrenic patients showed this "positive" response, while other, non-schizophrenic patients did not react to stimulation even when the stimulus intensity was very high. It seemed to us that the positive response clearly came from the knee: if the electrode lay in the cingulum, below or in front of the knee of the corpus callosum, the response was opposite, *i.e.*, the symptoms became worse.

Mesoloviotomy was effective in schizophrenic anxiety and tension, but without any effect in other psychiatric disorders. I used this approach on 48 patients, four of them under general anesthesia. A long-term postoperative follow-up study on these patients was published by Vilkki[16]. He found that mesoloviotomy had often been effective. An impaired visual imagery, similar to that seen after cingulotomy, was the only side-effect observed, but it could only be discovered in psychometric tests.

Subcaudate Tractotomy (Innominotomy)

Innominotomy has been extensively practised at the Brook General Hospital in London, England[7]. The disk-shaped bilateral lesions were aimed at undercutting the supraorbital cortex. The lesions lay at the antero-posterior level of the planum sphenoidale, extending from 6 mm to 18 mm from the midline and being 20 mm long in an antero-posterior direction. Two of my ten patients operated on in the subcaudate area had a "positive" emotional response to electrical stimulation, whereas most of them did not feel anything during stimulation. Subcaudate tractotomy is effective for chronic and recurrent depression, but less good for obsessive-compulsive neurosis. A serious complication of sexual disinhibition had occurred in 2% of the patients operated on by Knight. Minor side-effects had been observed in an additional 9% of the patients.

Very few innominotomies seem to have been done elsewhere. I have done it in ten patients. The result was positive in most of them. No side-effects were noticed. However, I do not like this approach, because the antero-posterior location of the lesions as related to the planum sphenoidale cannot be accurate. If the lesion lies too far posteriorly it may damage lenticulo-striate vessels. It also seems to me that today elderly patients with recurrent depression may be treated effectively with drugs, which is why they are not referred for surgery.

Postero-Medial Hpyothalamotomy

In 1962 Keijo Sano[13] reported on 6 erethistic, idiotic and epileptic patients who had had bilateral stereotactic lesions in the postero-medial part of the hypothalamus. The small spherical lesions, 4 mm in diameter, lay 2 mm posterior to the midcommissural point, 3.5 mm below the intercommissural line and 2.5 mm lateral to the wall of the third ventricle. In most patients the bilateral lesions resulted in a marked decrease of the sympathetic tonus and a diminution of the restless and aggressive behaviour. Sano, and after him others, have reported on large series of patients. They all had performed the operation at two sessions, with 1–3 weeks between the left and the right side.

In 1972, after having done postero-medial hypothalamotomy in two aggressive oligophrenic patients, I operated on a 23-year-old man who suffered from hebephrenic schizophrenia with extremely severe restlessness, agitation and autodestructiveness. Mentally the patient was already badly demented. Bilateral hypothalamotomy at one session had a very good calming effect. No side-effects were observed. When I examined the patient ten years later I observed that the effect had remained good. The patient was placid, pleasant and cooperative. He worked in the hospital garden. Because of this remarkable result, I operated in 1981 on another schizophrenic patient. Even this 53-year-old woman with intractable restlessness and agitation was demented, which was why no psychological tests could be done before operation. The calming effect was very good. Now, six years later, she works without problems in the kitchen of the mental hospital.

Since then I have operated on 10 restless, agitated, aggressive and destructive schizophrenic patients, five of them with CT guided technique, without ventriculography. Four of the patients had a normal intelligence and could be tested extensively before and after operation, which was always done in one session. The size of the bilateral lesions was 3 mm in diameter, which was confirmed by postoperative CT studies. Recently I operated on a 73-year-old man, who after a subarachnoid haemorrhage from an anterior communicating artery aneurysm seven years previously had developed a central allodynic and hyperaesthetic pain state with very severe restlessness.

Electrical high-frequency stimulation of the postero-medial hypothalamus causes a marked sympathicomimetic effect: bilateral pupillary dilatation, a gaze convergence inwards and downwards as seen in angry bulls, a rise of the blood pressure and heart rate, and a very marked rise in metabolism, measured from the pCO_2 of the expiratory air. The sympathicomimetic responses develop gradually within about 5–8 seconds after the beginning of stimulation and continue for a minute or two after it cessation. A thermolesion first on the left side causes a slight sympathicolytic effect, which increases when the other side has been coagulated.

The patients are only slightly drowsy after postero-medial hypothalamotomy, if the preoperative anxiolytic and sedative medication is reduced to a half of the previous level. They can usually be mobilized one day after operation. The operation is very effective for restlessness, aggressiveness and destructiveness, but in some of the patients the final effect does not appear immediately, but develops gradually two to six weeks after operation. In two patients the symptoms recurred within three weeks. One of them, a 73-year-old man with a previous aneurysm operation showed at the control CT study that the right-sided lesion lay in the right place, whereas the left sided lesion lay in the wall of the third ventricle. He has recently been reoperated and the short-term effect seems to be good. Left-sided electrical stimulation at the reoperation gave a strong sympathicomimetic response, but no stimulation response was obtained from the right side. Since the second operation the patient has been calm and cooperative without signs of adverse effects. The second patient with a recurrence of symptoms was also reoperated. Even in him the result seems to be good.

One patient who previously had had a bilateral cingulotomy, innominotomy and amygdalotomy elsewhere, became confused one week after the posteromedial hypothalamotomy. A slight confusion lasted for six months, after which her recovery has been remarkable. Now, three years later, she studies semantics and mathematics. I have not observed other side-effects. Most of my recent patients have undergone extensive psychometric tests before and after operation. More and more patients with severe agitation, restlessness and aggressiveness of schizophrenic origin are now being referred to us. I feel that postero-medial hypothalamotomy is one of the most rewarding psychosurgical interventions.

References

1. Ballantine HR Jr, Cassidy WL, Flanagan NB, Marino R Jr (1967) Stereotaxic anterior cingulotomy for neuropsychiatric illness and intractable pain. J Neurosurg 26: 488–495
2. Bingley T, Leksell L, Meyerson BA, Rylander G (1977) Long-term results of stereotactic anterior capsulotomy in chronic obsessive-compulsive neurosis. In: Sweet WH, Obrador S, Martin-Rodriguez JG (eds) Neurosurgical treatment in psychiatry, pain, and epilepsy. University Park Press, Baltimore, pp 287–299
3. Foltz EL, White LE (1962) Pain "relief" by frontal cingulumotomy. J Neurosurg 19: 89–100
4. Fulton JF (1952) The frontal lobes and human behaviour. Liverpool, Univ Liverpool Press
5. Herner T (1961) Treatment of mental disorders with frontal stereotaxic thermo-lesions. Acta Psychiat Neurol Scand 36 (Suppl 158): 1–140
6. Kelly D, Mitchell-Heggs N (1973) Stereotactic limbic leucotomy: a follow-up study of thirty patients. Postgrad Med J 49: 865–882
7. Knight G (1964) The orbital cortex as an objective in the surgical treatment of mental illness. The results of 450 cases of open operation and the development of the stereotactic approach. Br J Surg 51: 114–124
8. Kullberg G (1977) Differences in effect of capsulotomy and cingulotomy. In: Sweet WH, Obrador S, Martin-Rodriguez JG (eds) Neurosurgical treatment in psychiatry, pain, and epilepsy. University Park Press, Baltimore, pp 301–308
9. Laitinen LV (1972) Stereotactic lesions in the knee of the corpus callosum in the treatment of emotional disorders. Lancet 1: 472–475
10. Leksell L (1949) A stereotaxic apparatus for intracerebral surgery. Acta Chir Scand 99: 229–233
11. Meyerson BA, Bergström M, Greitz T (1979) Target localization in stereotactic capsulotomy with the aid of computed tomography. In: Hitchcock ER, Ballantine T, Meyerson BA (eds) Modern concepts in psychiatric surgery. Elsevier/North Holland, Amsterdam, pp 217–224
12. Narabayashi H, Nagao T, Saito Y, Yoshida M, Nagahato M (1963) Stereotaxic amygdalotomy for behaviour disorders. Arch Neurol 9: 1–16
13. Sano K (1962) Sedative neurosurgery. With special reference to postero-medial hypothalamotomy. Neurol Med Chir (Tokyo) 4: 112–142
14. Spiegel EA, Wycis HT, Marks M, Lee AJ (1947) Stereotaxic apparatus for operations on the human brain. Science 106: 349–350
15. Teuber H-L, Corkin SH, Twitchell TE (1977) Study of cingulotomy in man: a summary. In: Sweet WH, Obrador S, Martin-Rodriguez JG (eds) Neurosurgical treatment in psychiatry, pain, and epilepsy. University Park Press, Baltimore, pp 355–362
16. Vilkki J (1972) Late psychological and clinical effects of subrostral cingulotomy and anterior mesoloviotomy in psychiatric illness. In: Sweet WH, Obrador S, Martin-Rodriguez JG (eds) Neurosurgical treatment in psychiatry, pain, and epilepsy. University Park Press, Baltimore, pp 253–259

Correspondence: Dr. L. V. Laitinen, Rosendalsslingen, 21, S-18600 Vallentuna, Sweden.

Acta Neurochirurgica, Suppl. 44, 163–166 (1988)

Psychiatric and Neuropsychological Findings After Stereotactic Hypothalamotomy, in Cases of Extreme Sexual Aggressivity

G. Dieckmann*, B. Schneider-Jonietz, and H. Schneider****

* Department Functional Neurosurgery, Georg-August-University Medical School, Göttingen, Federal Republic of Germany, ** County-Hospital, Erlangen, Federal Republic of Germany

Summary

Report on 14 cases treated for aggressive sexual delinquency by unilateral ventromedial hypothalamotomy. Eight of them had a thorough psychiatric and psychological examination during follow-up.

The following constant modifications were found: Decrease in the domination by sexual drive, increase in the fluency in semantic contexts, increase of rapidity of visual image formation and of coordinative perception processes, positive modifications in the scope of some personality dimensions (poise, openness, self-criticism), decreased colour perception, increased appetite.

The structure of the patient's sexuality was not changed but the probability of a specific aggressive behaviour reduced, due to the diminished sexual drive thus allowing a positive consolidation of their social interactions with more harmonic relationships to the families or partners.

These results give proof that the many prejudices and aversions against this type of therapy are not justified.

Keywords: Sexual delinquence; stereotactic hypothalamotomy; postoperative psychiatric and psychological findings.

Introduction

The first report of stereotactic operations within the framework of sexual therapeutic measures was made in 1966 by Roeder in a case of paedophilia. Subsequently, this intervention was shown to cause a reduction in human libido and potency, whereupon Dieckmann and Hassler (1975) performed this operation in cases of aggressive sexual deviation. The particular moral and ethical problems connected with a therapy attempting to change specific behavioural patterns by means of a topistic intervention in definite brain structures accounts for the existence of a wall of prejudices, doubts, limitations and aversions against this type of therapy. Cries of "Deprivation of Free Will", "Modifications of Personality" and "Descent to Subhuman Levels" stood in the foreground of a lasting—extremely affectively and seldom objectively conducted—discussion.

The present paper is meant not only to give a report of the social outcome of the sexual delinquents on whom a medial anterior and ventromedial hypothalamotomy was performed between 1970 and 1979 (such a report was published in 1979), but also to present the so far unpublished findings of clinical psychiatric and psychological postoperative examinations of these delinquents (Schneider 1978, Schneider-Jonietz 1978). These findings disprove the sometimes too enthusiastically presented objections.

Material and Methods

From 1970 to 1979 we treated 14 patients with a unilateral ventromedial hypothalamotomy in the non-dominant hemisphere because of aggressive sexual delinquence. All of these patients had been living conflict dominated lives for years due to the peculiarity of their sexual problems. Various forms of their—at least partly—highly aggressive sexual delicts had resulted in great disorder in their way of life.

Eight of these 14 patients were thoroughly examined psychiatrically and psychologically after the intervention. The first postoperative examination occurred two to three weeks after the intervention. The second postoperative check-up took place one year after the operation. The schedule for this second examination could not be kept constant for all of the patients. Occasionally it was difficult to motivate the patients, who had by this time left the clinic, to come to such a check-up. This explains why only 8 of the 14 patients who had been operated on could be examined postoperatively.

A concept taking into consideration the clinical situation was used as the basis for the collection of our data; it agrees to a large extent with phenomenological descriptive models of earlier personality research. Gruhle (1922) considered the following factors as constant fundamental functions of the personality: activity; fundamental mood; susceptibility to strong emotions; will power; intrinsic relationships; environmental assimilation; self-acceptance. Allowing for such aspects, a classification system, in which the whole of the

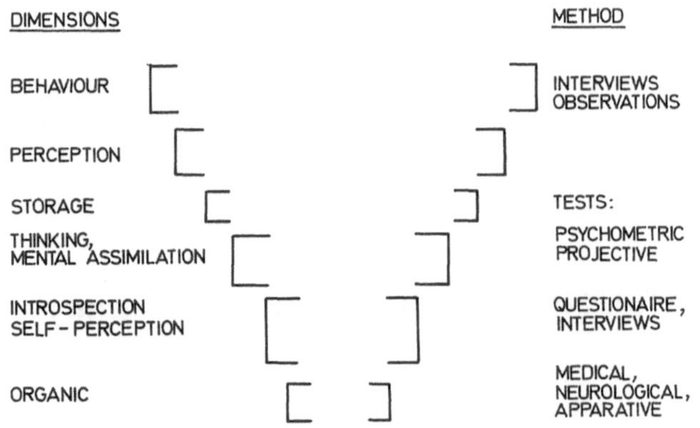

Fig. 1. Schematic illustration of the investigational methods and the personality structure which they determined

individual personality should be expressed, was developed. The following methods were employed to acquire data:

1. Physical examination techniques;

2. apparative procedures (EEG, registration of the micromuscle tone regulation);

3. external anamnesis and behavioural observation (psychiatric exploration, semistructured interview);

4. questionnaires (Freiburg personality inventory);

5. psychological tests, such as HAWIE, LPS, BENTON, d 2, GIESE, TAT, FPT.

All findings were coded and transferred to magnetic tape. For each patient 912 items were available for calculations. All data were processed statistically and analytically. The data gathered by apparative and psychological testing methods were mathematically tested for their randomness. For data with a normal distribution the T-test for independent samples from SPSSH-release 6.02 (statistical package for the social sciences) was used; for data whose distribution was not known the Man-Whitney U-test was employed.

Figure 1 illustrates schematically the investigational methods and the personality structure which they determined.

Results

One year after the operation all eight subjects examined postoperatively were found to have decreased sexual compulsion. With the decrease of their compulsion the severity of their aggressive sexual traits was greatly reduced. Three of the patients showed an accompanying decrease in their general initiative.

A marked consolidation of their social interactions was observed. Six of these patients had more harmonic relationships to their families and to their partners. Their occupational situation improved. Only one patient became more irritable and more labile after the operation.

Five of the eight patients exhibited an increased appetite followed by weight gain; this is a typical side-effect of hypothalamic ventromedial interruption.

Table 1 shows the postoperative results with respect to sexual behaviour, psychic and physical factors one year after the intervention.

Table 1. *Postoperative Results of Sexual Behaviour; Psychic and Physical Factors*

Factors	Numbers of patients (N 8)		
	Unchanged	Changed	
Sexuality			
Sexual compulsion	—	8/8	↓
Relationship to partner	—	8/8	↑
General initiative	3/8	5/8	↓
Aggressiveness	1/8	7/8	↓
Behavioural structure			
General	2/8	6/8	↑
In the family	2/8	6/8	↑
At work	3/8	5/8	↑
Side effects			
General psychic	7/8	1/8	
General physical	3/8	5/8	↓
Neurological findings	8/8	—	
EEG	8/8	—	

It is not possible to go into the exact details of the patients sexual modifications in this paper. However, Figure 2 shows a schematic diagram of one patient's sexual perceptions, which he wrote down after an unilateral hypothalamotomy. The diagram shows the oscillations of his sexual needs after the intervention. The

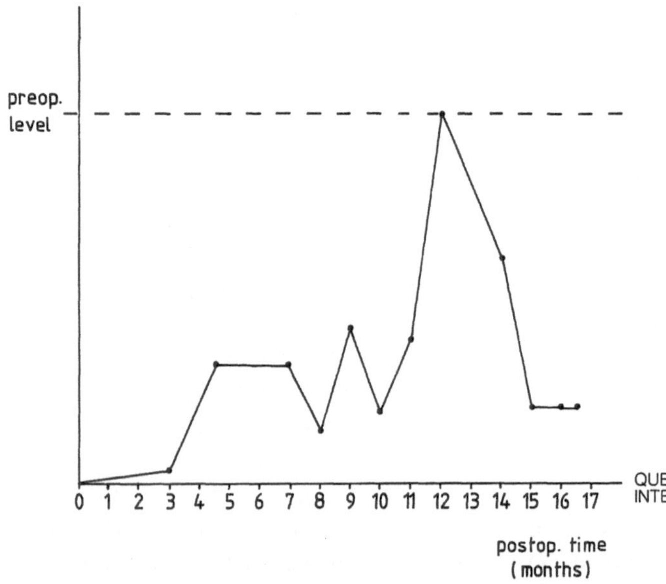

Fig. 2. Schematic diagram of a patient's sexual perceptions showing the oscillations of the sexual needs after the intervention. The dashed line shows the upper limit of sexual frequency before the intervention (from Schneider 1977)

dashed line shows the upper limit of sexual frequency, the so-called normal frequency before the intervention. The number of months after the intervention are given on the abscissa (from Schneider 1977).

Following a stereotactic ventromedial hypothalamotomy not only are the dynamic aspects of sexuality, such as compulsion and impulsitivity, diminished, but also organic components, which have to do with completion of the sex act. In contrast, the structure of the patient's sexual organisation remains unchanged: A paedophilic character, for example, is still retained. However, it is possible for the subject to adapt his sexual behaviour to the specific conceptions and expectations of our society. Moreover, cognitive assimilation components become just as important as emotional aspects for the patients full enjoyment of their sexuality.

Disconnections in the ventromedial and anterior hypothalamus imply an impulse reduction in subcortical structures. Such a reduction could result in a modification of the thalamic integration pattern. We do not want to attempt to establish a psychosomatic parallelism in this connection; however it should be noted that the changes in the thalamic integration pattern could be a reason for observable changes in psychic functional regions having no phenomenological connection with sexuality.

Thus, in addition to the described reduction of sexual compulsion, the following constant changes were found to occur as direct results of the operation:

1. An increase in semantic fluency;
2. an increase in the speed of visual image formation and coordinative processes in mental and visual perception;
3. a modified ability to perceive colours.

These changes (Table 2) are especially clear, *e.g.*, in the assimilation of complex perception stimulii. In the case of the projective technique TAT this is expressed in a paucity of words in the patients interpretations. In comparison to pre-operative examinations these interpretations are more attuned to the specific visual stimulus. Affective and egocentric components become less important after the operation.

The test's findings were confirmed by the subjective statements of the patients themselves. Compared to

Table 2. *Constant Modifications After Unilateral Ventromedial Hypothalamotomy*

1. Decrease in the domination by sexual drive
2. Increase in the fluency in semantic contexts
3. Rapidity of visual image formation and of coordinative perception processes
4. Positive Modifications in the scope of the personality dimensions (poise, openness, self-criticism)
5. Decreased colour perception
6. Increased feeling of hunger

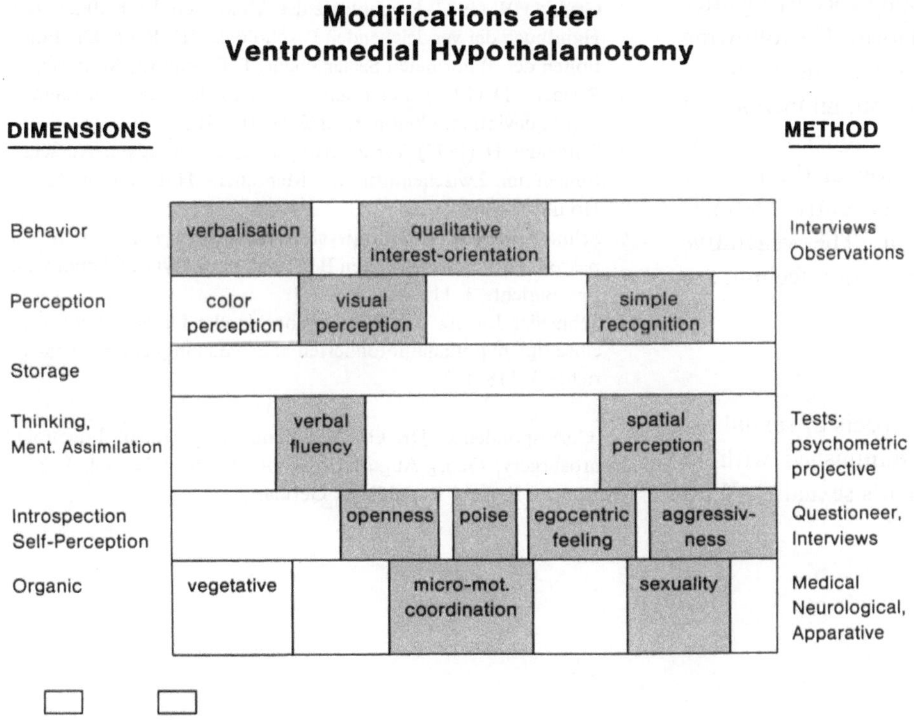

Fig. 3. Psychic changes after ventromedial hypothalamotomy. Items hatched (modified from Schneider 1977)

their previous behaviour, they seemed less self-centred in their thoughts and actions. This tendency dissimulate their socially less desired peculiarities (such as aggressivity and impulsivity) which was clearly improved. The questionnaire accompanying the second postoperative examination showed a significant change in their self-attunement: They have become increasingly self-critical and more open.

Several patients stated that they felt a certain loss of initiative. It was however possible for them to compensate for this perceived lack of initiative by using their will power.

All of the patients consider themselves to be more poised and relaxed. They appear to be free from the pressure of compulsive sexuality.

If a hypothalamic intervention has effects on the range of colour perception, this demonstrates the neuroanatomical connection of the hypothalamus to the optic nerves, to the hypothalamic optic root. This could be interpreted as showing that hypothalamic impulses have an activating effect on the perceptional process.

The unilateral hypothalamic intervention can also cause flawed control of vegetative functions. This is shown in a decreased satiation feeling in five of the eight patients. It can be explained neurophysiologically by the proximity of the stereotactic interrupted ventromedial nucleus to the lateral nucleus of the hypothalamus.

Figure 3 presents the psychic changes occurring after such an intervention in a simplified form. The following changes were noted: An increase in the fluency of semantic processes and visual perception; an increase of self-criticism in their self-judgement; inner harmony, which allows more rapid mobilization of the psychic abilities; a reduction of affective egocentric feelings; modification of colour perception. The vegetative changes, *i.e.* modifications of the satiation feeling, are also shown.

Discussion

The intervention allows dynamic aspects of sexual experiencing and behaviour to be diminished without changing the structure of the patient's sexuality. With this dynamic modification the probability that a specific aggressive behaviour will occur is reduced. Due to a relieving of intrinsic urges, other psychic abilities, which are important for a successful shaping of one's life, can be positively influenced.

On the other hand, if a sexual symptom complex manifests itself merely in impulsive fantasies, which cause the person to suffer but are not acted upon, a psychosurgical intervention is not indicated.

It has been shown that this psychosurgical intervention neither prevents the patient from deciding how he wants to live nor does it hinder his self-development. The intervention creates a modified inner milieu that he can definitively shape according to his primary abilities. The group of sexual delinquents investigated here has supplied the proof. That, in spite of this, the psychic and organic make-up of the patient is of great importance for the success of this intervention goes without saying. Interventions on an already damage brain lead to side-effects.

References

1. Dieckmann G, Hassler R (1975) Unilateral hypothalamotomy in sexual delinquents. Proc 6th Symp Int Soc Res Stereoencephalotomy, Tokyo 1973, part II. Confin Neurol 37: 177–186
2. Dieckmann G, Horn HJ, Schneider H (1979) Long-term results of anterior hypothalamotomy in sexual offences. In: Hitchcock ER, Ballantine HT Jr, Meyerson BA (eds) Modern concepts in psychiatric surgery. Elsevier Biomedical Press, pp 187–195
3. Gruhle HW (1922) Psychologie des Abnormen. In: Kafka G (ed) Handbuch der vergleichenden Psychologie. III. Band: Die Funktionen des abnormalen Seelenlebens. E Reinhardt, München
4. Roeder FD (1966) Stereotaxic lesion of the tuber cinereum in sexual deviation. Confin Neurol 27: 162–163
5. Schneider H (1977) Veränderungen nach operativen Ausschaltungen im Zwischenhirn des Menschen. Habilitation Thesis, Homburg-Saar
6. Schneider H (1978) Neuropsychologische und psychiatrische Aspekte bei der Bewertung von Befunden nach Hypothalamotomie. bga-Berichte 3: 115–117
7. Schneider-Jonietz B (1978) Psychologische Untersuchungen an einseitig hypothalamotomierten Sexualdelinquenten. bga-Berichte 3: 118–122

Correspondence: Dr. G. Dieckmann, Department Functional Neurosurgery, Georg-August-University Medical School, D-3400 Göttingen, Federal Republic of Germany.

Acta Neurochirurgica, Suppl. 44, 167–169 (1988)

Experiences in Psycho-Surgery in the Netherlands

J. van Manen[1] and **C. W. M. van Veelen**[2]

[1] Department of Neurology, Academisch Medisch Centrum, Amsterdam Zuidoost, The Netherlands, [2] Department of Neurosurgery, Academisch Ziekenhuis, Utrecht, The Netherlands

Summary

The authors report on their experiences in 54 cases operated upon for various psychiatric diseases including compulsive neurosis, depression, anxiety, tension and in some of this group also automutilation; intractable temporal lobe epilepsy and aggressive behaviour; aggressive behaviour and minor epileptic problems; severe mental retardation, restlessness, automultilation and in some of this group also aggression.

Operative procedures have been fronto-basal lesions according to Knight and Bridges, as well as lesions in the cingulum, the paracingular white matter, the anterior part of the radiation of the corpus callosum and the basal frontal region, using the technique of Crow. Amygdalotomy and thalamotomy was performed for epilepsy, aggression and automutilation in the mentally retarded patients.

Because of the small number of patients and the variety of different diseases and techniques no statistically valid analysis of the results is possible.

In 1975 a number of Belgian and Dutch neurosurgeons, neurologists and epileptologists formed a workgroup to study the value of surgery in severe intractable psychiatric disorders and epilepsy. Within the scope of this workgroup, a total of 54 Dutch patients have undergone psychosurgical operations for various indications;

— compulsive neurosis—22 cases,

— depression, anxiety, tension and in some patients also automutilation—nine cases,

— intractable temporal lobe epilepsy and aggressive behaviour, mostly combined with mental retardation—eight cases,

— aggressive behaviour, and minor epileptic problems—two cases,

— severe mental retardation, restlessness,

Table 1

	No.	Basal front.	Cing. basal front.	Amygdala 1×	Amygdala 2×	Fornix	L. M. I.	Leucot.	Temp. lob.	Improved, much improved	Slightly improved	Not improved
Compulsive neurosis	22	12	10							11		10+1 †
Depression, anxiety, tension, automutilation	9	6	4							3	4	2
Temporal lobe, epilepsy aggression	8			2	6	3			1	1	2	5
Aggressive behaviour	2			2						2		
Automutilation mental retardation, restlessness	13				1		12	2		4	7	1+1?
	54	18	14	4	7	3	12	2	1	21	13	20

Basal. front. = Basal frontal lesion according to Knight.
Cing. basal front. = Cingular and basal frontal lesions according to Crow.
L. M. I. = lamina medullaris interna thalami.
† Deceased.
? Short follow up.

— automutilation, in some patients also aggression—13 cases (Table 1).

The results in the group of 22 patients with compulsive neurosis will soon be reported in detail by Cosijns, Haaijman and Ceha.

In 12 of these patients a basal frontal lesion was performed according to the technique of G. Knight and Bridges. The other ten were treated according to the technique of Crow, with multiple electrodes and lesions in the cingulum, the para-cingular white matter, the anterior part of the radiation of the corpus callosum and the basal frontal region. Half of these patients were reported to show improvement. No deterioration in the mental state was reported. Pre- and post-operative behaviour therapy was essential for optimal results.

The results of the acute technique of Knight were slightly superior to those with the chronically implanted electrodes as used by Crow. One compulsive patient treated according to the method of Crow died in the period that the coagulations were being made, because of an accident unrelated with the operation.

Of the second group of patients with depression, anxiety, tension or automutilation, four were operated according to Crow and two of them improved slightly and one more definitely. One patient with compulsive behaviour had a second operation according to Knight, after an unsuccessful Crow procedure. However both operations were without permanent relief. Six patients had orbitofrontal lesions according to Knight including the one above mentioned. Two of them improved clearly, namely with one tension and anxiety and one with an urge to automutilation. One patient with anxiety, tension and depression improved slightly and one with depressions and compulsive thoughts showed no improvement. The sixth patient with depression, compulsive traits, anxiety and automutilation was treated with the Knight operation and improved slightly except for the automutilation. The follow-up of all 31 mentioned patients was 4–10 years and more. In the treatment of depression our results were not satisfactory in this small number of patients.

From the group of eight patients with intractable epilepsy, automutilation and aggression only three cases improved regarding their behaviour and in no case the epileptic problems were influenced in a positive or negative way. No patients deteriorated as regards their mental state. In six of these patients bilateral amygdaloid lesions were performed and in three cases also a unilateral fornicotomy. In one case a unilateral lesion was made in the amygdala and on the other side a temporal lobectomy was performed.

One patient with only a unilateral amygdalar lesion improved most clearly. Follow-up was 1–3 years. The two cases of aggressive behaviour and minor epileptic problems had good results as regards aggression with a unilateral amygdalar lesion. In one case aggression was greatly reduced and in another aggressive behaviour ceased completely. Both patients had a follow-up of ten years. Pre-operatively the amygdala showed an atrophic lesion in the CT scan.

So there are three cases with a more or less clear pathology, mainly in one amygdaloid nucleus that improved after a unilateral stereotactic lesion on that side. This is in contrast to the seven amygdaloid lesions in more bilateral pathologic cases.

There were 13 cases of automutilation, mental retardation, and restlessness, and in some cases aggressive behaviour. In 12 of these cases a lesion in the internal medullary lamina of the thalamus bilaterally was made.

The location of these lesions was chosen according to the description of Hassler and Dieckmann. Two patients had to be operated for a second time in the same location.

In four cases a useful improvement was obtained. Four cases improved slightly and two cases did not improve. Slight improvement meant that the patient was better to handle for nursing. The two not improved patients were operated on for a second time. In one case a classical leucotomy was made and the patient slightly improved. In another case a bilateral amygdalotomy was made without result. From one case no follow-up could be obtained and one case is only slightly improved after an operation some months ago. From the other cases the follow-up is more than one year, but in most cases 5–10 years.

The 13th case was treated by a classical leucotomy and improved slightly.

Regarding the placement of the lesion some remarks can be made. An accurate placing of the lesion according to Knight in the orbitofrontal area is difficult to obtain. The position of the tuberculum sellae or limbus sphenoidale in relation to the orbitofrontal white matter is uncertain. The space between the caudate and accumbens nucleus and the orbitofrontal grey matter is very narrow. We have one autopsy case in which the orbitofrontal grey matter was lesioned on several locations. The desired width of the lesion in the horizontal plane is difficult to determine and in some cases the lesion was enlarged in the lateral direction in a second operation.

We have no personal experience with acute cingular lesions but from the literature we had the impression that most of these lesions were made far laterally, thus not only in the cingular gyrus.

With the Crow method by means of electrostimulation a discrimination could be made between location of the electrodes in white or grey matter. Thus except in the cingulum all lesions were restricted to the white matter.

Only in one patient electrostimulation affected the behaviour. In this patient suffering from obsessive compulsive behaviour the abnormal behaviour was enhanced by stimulation in the cingulum. In general the best effects were obtained when the lesioning was started in the orbitofrontal white matter. In these cases the total extent of the lesion could be restricted to ± 1 cm^3. Extending the lesion had no further beneficial effect.

Capsular lesions are difficult to restrict to the narrow anterior limb of the internal capsule and the caudate and putamen will be easily damaged.

We did not make hypothalamic lesions because it is our opinion that in that region there are too many intermingled systems to perform selective coagulations. Generally speaking it is difficult to restrict a lesion to a limited number of anatomical circuits when the locations is chosen in regions of the brain as deep as the subthalamus, hypothalamus and brain stem. For the interpretation of the effects of lesions in psychosurgical interventions it is of course more reliable when a restricted number of anatomic and physiologic circuits is involved.

In the literature there is some discussion about the question whether a lateral or medial amygdaloid lesion should be made. We made the lesions in the middle of the nucleus as far as possible. The variation of the nucleus in the mediolateral direction appeared large judged on the air encephalogram. Therefore we thought it more safe to place the lesion medially in the nucleus.

The lesions in the internal medullary lamina of the thalamus are problematic. A part of the nucleus is located in front of the mamillothalamic tract and a larger part behind that tract. It is difficult to determine a direction of the electrodes that coincides with a large part of the nucleus. In our opinion it is also difficult not to damage the mamillothalamic tract and certainly this will have happened in some cases. The vascular supply of the area depends on the polar thalamic artery which is running approximately parallel to the mamillothalamic tract and this artery can easily be damaged. Therefore these lesions are dangerous and can only be made in cases of severe mental retardation. Perhaps the development of amnesia caused by a lesion of the mammillothalamic tract is the reason for success in these cases of automutilation. These lesions were also advocated by Hassler in the Gilles de la Tourette syndrome and for compulsive neurosis but we never dared to make a lesion in that region for these indications.

In conclusion we think that psychosurgical interventions are only indicated in cases that appeared intractable to all other forms of therapy. Psychosurgery can be helpful in cases of compulsive neurosis, anxiety, tension, aggression and automutilation (see Table 1).

Our experience with depression is limited but not very encouraging. In automutilation, mental retardation, and restlessness intralaminary thalamic lesions can give satisfactory results. In aggressive disorders without mental retardation amygdalotomy can give good results if there is a clear pathology in one amygdaloid nucleus. Postoperative psychiatric care is of great importance for obtaining worth-while results.

Correspondence: J. van Manen, M.D., Department of Neurology, Academisch Medisch Centrum, Meibergdreef 9, 1105 AZ Amsterdam Zuidoost, The Netherlands.

Acta Neurochirurgica, Suppl. 44, 170–172 (1988)

Psychosurgery and Personality—Some Legal Considerations

H. Nys

Katholieke Universiteit Leuven, Seminarie voor Gezondheidsrecht, Rechtsfaculteit, Leuven, Belgium

Summary

The practice of psychosurgery is subjected to different rules, sometimes written, often unwritten, in the different jurisdictions.

No commonly accepted definition on what psychosurgery exactly is has been given by the national legislatures. Some legislatures have simply outlawed psychosurgery; some have no specific rules; others have provided a stable legal basis to it, while protecting the patient (informed consent; review committees; second opinions).

Keywords: Psychosurgery; regulation; informed consent; Western Europe; United States of America; Australia; Soviet Union.

Introduction—Analytical Problems

Belgian lawyers, living and working on the crossroads of Europe and in a multicultural and multilingual country like Belgium are used to dealing with foreign law systems. The practice of medicine in general and of psychosurgery in particular is subjected to different rules, sometimes written, often unwritten, in the different jurisdictions. Even if the hypothesis is true that there are relatively few obstacles to prevent legal concepts as "consent" and "fault" being transplanted across boundaries (Shaw 1986, 865) it would be impossible and indeed annoying giving even a short survey of the legal state of affairs regarding psychosurgery in the different countries in the Eurasian Region.

During the preparation of this paper, the author has been confronted with another analytical problem, namely the diverging opinions on what psychosurgery exactly is. For lay-people it is somehow troubling to discover that no commonly accepted definition of psychosurgery seems to exist. It is interesting to have a brief look at the way some national legislatures have dealt with this problem. Some have given a rather broad definition of psychosurgery. For instance, section 57 of the English Mental Health Act 1983 refers to "any surgical operation for destroying brain tissue or destroying the functioning of brain tissue", and the Ontario Mental Health Act defines the performance of psychosurgery as any procedure that, by direct or indirect access to the brain, removes, destroys or interrupts the continuity of histologically normal brain tissue or which inserts indwelling electrodes (Gordon and Verdun-Jones 1983, 64–65). In other acts, one founds a very limited enumeration of procedures. For instance, section 19 of the Mental Health Act 1976–1977 of South Australia defines psychosurgery by mentioning six specific procedures. Still other legislations do not give any definition at all, for instance the Nova Scotia Hospitals Act (idem). Actually, psychosurgery is not a single thing but the name given to a somewhat heterogeneous assemblage of procedures (Kleinig 1985, 4) designed to alter disordered mental states and/or behaviour by removing, destroying or severing apparently healthy brain tissue. Each procedure has to be judged on its own merits and disadvantages, also legally. Lawyers should avoid an overall judgment on psychosurgery as such.

Preoccupation of Medical Law with Psychosurgery

Why is medical law preoccupied with psychosurgery? The answer could be because psychosurgery has to do with the seat of personality, *i.e.* the human brain. However, a better question to begin with is the following: is medical law really preoccupied with psychosurgery? Compared to the abundant Anglo-Saxon legal literature on psychosurgery, it is striking to see that in continental Europe psychosurgery has attracted the attention of legal scholars much less. This apparently lack of interest can be interpreted in different senses—it can result out of the fact that psychosurgery is simply considered as illegal or at least at the margins of legality on one hand. Another explanation could be that psychosurgery is considered as more or less an accepted medical procedure, that poses no special legal problems. The first interpretation should not be underestimated. Indeed, psychosurgery was outlawed in the Soviet Union in the early fifties. Although there are no western European countries that have taken the same position—in the early eighthies some voices have pleaded for a prohibition of psychosurgery on minors in Switzerland—(Gsell, 1984, 241) it remains in some European countries a procedure difficult to bring in line with the requirements of medical law. This is especially true in the so-called Latin countries of Southern Europe but also in France and Belgium, where the patient's consent to a medical procedure is less preponderant than in the more Germanised countries such as the Federal Republic, the Netherlands and Scandinavia. In Belgium for instance, the destruction of histologically normal brain tissue could be considered as battery and even the explicit consent of the patient would not guarantee immunity from prosecution to the surgeon. Not the consent of the patient but the therapeutical purpose is the ultimate cause of justification for a physician invading the bodily integrity of his patient. Whether or not such a therapeutical purpose is present in a psychosurgical procedure is a very lively debated question and a judge confronted with it would certainly receive diverging expert opinions. Fortunately for psychosurgeons, there is an implicit assumption in Belgian law that any medical act has at least some therapeutic value and it is up to the prosecutor to ascertain that no such value at all has been present. This is probably one of the reasons why there are no precedents on psychosurgery in Belgian jurisprudence.

Nonetheless, psychosurgery remains a debatable procedure and one may wonder whether in these countries a legal intervention would not be welcome, not to outlaw psychosurgery but to provide a stable legal basis for it.

The Consent of the Patient

Although the patient's consent is not in itself enough to justify psychosurgical intervention, it is understandably a necessary condition to it. Apparently, the prerequisite of consent is generally accepted, but the way it is brought in practice may differ widely. Or to put it in another way, the right to consent is not at issue because it is generally accepted that for a doctor to touch a patient's body is unlawful without consent. What is at issue is precisely what doctors must do to facilitate the giving of that consent in circumstances where the patient actually understands what the issues are (Shaw 1986, 867). In other words the issue is whether or not the consent of the patient should be informed. As a legal doctrine informed consent originated in the United States and spread in a modified form to Canada. In a recent comparative study on informed consent, Shaw concludes that informed consent, as a legal doctrine, has failed so far to have any significant impact on the English legal system. This conclusion is also valid for other European jurisdictions, *e.g.* the Belgian. Under Belgian jursiprudence a physician is required to inform his patient so that he is able to refuse a medical intervention. Thus, the patients has a right to refuse a proposed procedure. Because of this, we should speak of informed refusal instead of informed consent. On the other hand the doctrine of informed consent is well developed in the Federal Republic of Germany. An important reason for this is historical and much of the impetus towards the development of concrete protection for individual self-determination comes from the experiences of Germany under the Third Reich. The German doctor's duty of disclosure also originates from the human rights provisions of the Constitution and likewise it is a reaction to the lack of respect for human dignity in Hitler's Germany. The obligation of the German physician to inform his patients is governed by patient-centred standards requiring them to tailor their disclosure to the needs of the individual. Emphasis is placed on active comprehension, rather than the passive reception of knowledge by the patient. The picture is still more complicated with regard to psychosurgery because the capacity of the patient to understand the information might be very limited.

Regulation of Psychosurgery

In an effort to protect these patients different legal strategies have been elaborated. One of them is the banning of psychosurgery or imposing such strict conditions that psychosurgery is effectively killed as in Oregon and California (Grimm 1980, 435). An interesting question is whether a state has a right to forbid the practice of psychosurgery on its territory. Both the right to health care and the right to privacy should be carefully balanced against a state's right to protect vulnerable patients.

Another strategy consists of shifting the requirement of informed consent from the patient to someone else, or making the validity of the informed consent of the patient dependent upon the approval or review of someone else. For instance, The New South Wales Mental Health Act of 1983 states that psychosurgery may not be performed upon a voluntary patient without his/her informed consent and without the approval of a psychosurgery review board (and in some cases, the Supreme Court). The psychosurgery review board consists of seven members including a legally qualified chairperson, a neurosurgeon, a neurologist, a clinical psychologist, a person nominated by the New South Wales Council for Civil Liberties and two psychiatrists. The membership of the council is thus clearly dominated by health professionals but the act specifies the requirements for informed consent in an unusual degree of detail (Verdun-Jones 1986, 108–109). The English Mental Health 1983 is very complex with regard to the consent to treatment. It distinguishes three categories of treatment—some treatment that requires the patient's consent *and* a second opinion, some that requires the patient's consent *or* a second opinion and, finally, treatment where consent is not required. Some forms of treatment are specified in the Act. For instance psychosurgery falls under the first category and thus consent and a second opinion are required. Electro-convulsive therapy (ECT) belongs to the second category. For treatments that belong to the first category two people who are not doctors are appointed by the Mental Health Act Commission to participate in the consent procedure and they have to certify that the patient is capable of consenting and has consented. To be capable of consent, a patient must be able to understand the nature, purpose, and likely effects of the treatment in question. The decision as to whether a patient is capable of consent is made by an independent doctor and the two other people appointed by the Mental Health Act Commission. The act also contains important provisions concerning emergency treatment which enable psychiatrists to administer treatment without considering the issue of consent. One author comments on these provisions as follows: "The effect may very well be that psychosurgery will no longer be possible, a result which would certainly not be universally regretted" (Hoggett 1984, 210). We should be very prudent in making such statements because they seem to overestimate the regulatory force of legislation. I am inclined to agree with a commentator of the Oregon and California acts, namely that "where bureaucratic and legislative intrusion with a specific medical practice is successful, the practice is already moribund" (Grimm 1980, 436). Comparable examples of recent legislative reforms in Commonwealth jurisdictions to establish some degree of patient control over treatment decisions could be given. For instance, in South Australia the Mental Health Act requires the authorisation of three psychiatrists before psychosurgery is employed, also the patient's consent is required providing he or she has "sufficient command of his or her mental faculties to make a rational judgment". In the absence of such "command" on the part of the patient, the consent of a guardian or nearest relative will suffice.

Regulation in Continental Europe—Conclusion

However in continental Europe much less legislative reform, if any, has taken place. This should not be misunderstood because both the mental health legislation in general and the rules governing medical malpractice may offer a legal framework within which psychosurgery should find and respect its place. Moreover, the regulation of informed consent is only one aspect, albeit an important one, of this legal framework. Other components that have received less attention but are nevertheless of equal importance are the indications for psychosurgical procedures and the professional standard that has to be attained by those practizing psychosurgery.

References

1. Hamilton JR (1983) Observations on the mental health act 1983. Int J Law Psychiatry, pp 371–380
2. Hoggett B (1984) Mental health law. 2nd ed. Sweet and Maxwell, London
3. Gordon R, Verdun-Jones SN (1983) The right to refuse treatment: commonwealth developments and issues. Int J Law Psychiatry, pp 57–73
4. Goudsmit W (1978) Opmerkingen over psychochirurgie. Delikt en delinkwent 4: 228 (Remarks on psychosurgery)
5. Grimm RJ (1980) Regulation of psychosurgery. In: Valenstein ES (ed) The psychosurgery debate. Scientific, legal and ethical perspectives. Freeman, San Francisco, pp 421–438
6. Gsell O (1982) Ethique et morale dans les traitements corporels en psychiatrie: thérapie electroconvulsive, psychochirurgie. Proceedings of the 6th World Congress on Medical Law, I, Gent, p 236
7. Kleinig J (1985) Ethical issues in psychosurgery. George Allan and Unwin, London
8. Leenen HJJ (1981) Gezondheidzorg en recht. Samsom, Brussel, Alphen a. d. Rijn, pp 145–147 (Health care and the law)
9. Shaw J (1986) Informed consent: a German lesson. International and Comparative Law Quarterely, pp 864–890
10. Verdun-Jones SN (1986) The dawn of a "New Legalism" in Australia? The New South Wales Mental Health Act, 1983 and related legislation. Int J Law Psychiatry, pp 95–118

Correspondence: Prof. Dr. H. Nys, Katholieke Universiteit Leuven, Seminarie voor Gezondheidsrecht, Rechtsfaculteit, Tiensestraat 41, B-3000 Leuven, Belgium.

Acta Neurochirurgica, Suppl. 44, 173–178 (1988)

Ethics of Psychosurgery

A. J. Bouckoms

Harvard Medical School, Massachusetts General Hospital, Boston, Massachusetts, U.S.A.

Summary

The ethics of psychosurgery involve questions of moral philosophy and pragmatism in alleviating human suffering. The weighing of scientific data along with philosophical oughts and shoulds is required. The medical literature indicates definite efficacy for some kinds of limbic surgery, mainly cingulotomy and capsulotomy, in some kinds of conditions, namely major depression, pain and anxiety. The relative utility of these procedures given the severity of the illnesses and the safety of the procedures described is significant. Ethical and moral conflicts over altruism, autonomy and suffering require recognition before their due considerations (Kleinig 1985). The following recommendations emerge from these considerations:

1. No consideration of ethics in psychosurgery is complete without consideration of both the scientific data and moral conflicts.
2. The considerable efficacy and safety of cingulotomy and capsulotomy must be acknowledged.
3. Indications and contraindications do exist for selecting patients. Major psychiatric Axis I diagnoses of depression and anxiety are the indications. Personality disorders are not indications.
4. Peer review, unfettered consent and knowledge of the psychodynamics of severe illness are three ingredients necessary for wise decisions about performing limbic surgery.
5. The liberal advocation of autonomy without responsibility is an amoral, not liberating, point of view.
6. Politics should be denounced as the most serious ethical problem in medical decision making. Political intrusion into the scientific matters and the doctor-patient relationship has created ethical problems with psychosurgery and continues to do so today.

Keywords: Psychosurgery; cingulotomy; ethics.

Introduction

The word ethics is defined in two ways (Oxford English, Websters, Larousse Dictionaries). One meaning is a standard of conduct; as derived from the Greek "ethos" meaning custom or characteristic spirit and belief of the community. The second meaning is from the Greek root "ethikos" meaning moral philosophy. This second meaning defines ethics as a system or code of morals or principles adopted by a particular philosopher, religion, or group. These two meanings of the word ethical: namely standard conduct in the community or a moral philosophy, have significantly different connotations. The first meaning is a naturalistic one, based on what is observed and what is done in the community. These norms of behaviour become standards of conduct based on pragmatism. The second connotation of ethics as a moral philosophy imputes hypothetical or adjunctive thinking where ideals and concepts are most important. Ideas about what should, ought, or could be done are the currency of this moral-theological definition of ethics. Ethics as a set of should, oughts, and coulds is intellectually and practically quite different from the pragmatist's world of observations and acts. The primary focus of this discussion will be on the ethical problems of psychosurgery as pragmatic problems. These are the common ethical problems of dealing with any kind of sick person, be it cancer, chronic illness, or the mentally impaired. These pragmatic problems concern the efficacy, relative utility, and altruistic exploitative aspects of the surgery. The ethics of psychosurgery as problems of moral philosophy will be discussed later as questions of autonomy, nobility and politics. The obstruction of certain moral positions to the optimal care of patients in the real life clinical setting will be described.

Ethics as Questions: Pragmatic and Moral

Ethical problems in clinical medicine are questions that arise from disparate information or points of view. An ethical question in medicine arises when it is not clear what is best for the patient, or where what is best for the patient is compromized. Ethical conflicts may arise from problems in six areas.

1. Uncertain Efficacy

More than 19 different kinds of central neurosurgical procedure have been described for the relief of intractable psychiatric disease and pain (Bouckoms 1984). Procedures range from early non-selective ablative procedures, to modified procedures, and modern limited stereotactic operations. Therefore, it is impossible to make any general statement about the efficacy of psychosurgery. The four procedures most often used are stereotactic cingulotomy, subcaudate tractotomy, capsulotomy and thalamic lesions. I will limit my discussion on efficacy to these four procedures. The other procedures have fallen into disuse because of the significant morbidity associated with them (Sweet 1973). I shall summarize efficacy with four recent reviews on the efficacy of cingulotomy, subcaudate tractotomy, capsulotomy and thalamic lesions.

Cingulotomy

Ballantine recently reported a 20-year experience with 198 psychiatric patients who were treated with bilateral cingulotomy (Ballantine 1987). This is the largest series of patients described in the literature, followed for an average of 11 years, 26 patients (13%) were fully recovered and stable without recourse to ongoing psychiatric treatment. 46 patients (23%) continue to need psychiatric supervision and medication, but otherwise are functioning normally. An additional 51 patients (26%) have varying degrees of psychiatric disability, requiring ongoing supervision and treatment. Nevertheless, these patients are markedly improved over their pre-operative status. If one compares those patients with affective disorder to those with schizophrenia and personality disorders, the superiority of the treatment for the affectively disordered group is notable. 49 of 120 affectively ill patients achieved very good results, compared to 8 out of 32 obsessives and 7 out of 14 generalized anxiety patients. The absence of operative mortality and the low incidence of serious complications (0.03% hemiplegias and 1% seizure disorders) in a series of 696 bilateral cingulotomies represent firm evidence of the safety of this procedure.

Subcaudate Tractotomy

Psychosurgical lesions of the ventro-medial quadrants of the frontal lobes have been widely used for the treatment of affective illness. Strom-Olsen (1971) reported on 210 psychiatric patients treated with this procedure. They have found two-thirds of the patients had less depression, and one-third to two-thirds had significantly improve anxiety. Mortality was less than 1%, epilepsy less than 2%, and behavioural problems occurred in 3%. Similar positive results have been found in Shevitz' review in 1976 where he reported that 93% of patients treated with subcaudate lesions suffered no undesirable side effects.

Thalamic Lesions

A review of thalamic lesions for pain by Sweet (1980) showed the efficacy of these lesions in the short term with reduced efficacy at follow-up. However, this procedure has significant side effects, making it compare unfavourably with other psychosurgical procedures. For these reasons, it is less common having a less favourable risk-benefit ratio. In conclusion, the large number of patients treated by a variety of neurosurgeons with sub-caudate, cingulate, and thalamic lesions over more than 20 years has established the efficacy of these procedures for certain diagnoses in certain well-controlled circumstances.

Anterior Capsulotomy

Mindus *et al.* (1987) have contributed the largest series to the more than 300 cases of affective disorder treated with stereotactic interruption of the anterior fibers of the internal capsule (anterior capsulotomy). Significant improvement has occurred in two-thirds of patients. These people have had generalized anxiety, phobic anxiety, or obsessive anxiety for a mean of 15 years. Improvement of Axis V of the DSM III was from a mean of 6.1 preoperatively to 4.4 postoperatively. Carefuly examination with psychological tests showed a restoration of personality to normal with less somatic and psychic anxiety. Some post-operative apathy was noted but this gradually abated over 2 to 3 months. Preliminary data suggests that capsulotomy is particularly effective in anxiety disorders relative to cingulotomy which is primarily effective in depression.

2. Relative Utility—the Moral Cost

Ethical problems may arise when the relative utility of the specific psychosurgical treatment compared to other treatments for the condition is in doubt. The question of relative utility begs the questions of indications for psychosurgery per se, the specific inclusion-exclusion criteria, and the adverse effects of these procedures. To our knowledge there has been no effort to define specifically what the indications are for psychosurgery. Intractable suffering with anxiety, depression, and pain have been the general but loosely defined guidelines. At Massachusetts General Hospital, Bouckoms, Cassem, Ballantine, and Murray have defined a set of specific criteria that form the guidelines for the

Table 1. *Selection Criteria for Limbic Surgery*

1. Psychiatric diagnosis	DSM III/III-R major affective disorder anxiety disorder
2. Pain	with or without psychiatric diagnosis
3. Disability	poor to grossly impaired (DSM III 5–7)
4. Intractability	pharmacotherapy electrical stimulation psychotherapy
5. Informed consent	the patient must consent
6. Treating professionals	tertiary hospital with consensus of neurosurgery, psychiatry and neurology
7. Evidence	written past history family history
8. Exclusion	delusional psychosis somatoform disorders 1° personality disorder 1° substance abuse criminality/poor cooperation compensation neurosis, malingering unwilling to have psychiatric care

procedure of cingulotomy (Table 1). These guidelines require either specific major affective disorder or anxiety disorder, intractable pain, with or without psychiatric disorder, and a clear pattern of disability, intractability, absence of other psychopathology and a multidisciplinary approach. Informed consent from the patient is an absolute prerequisite for any psychosurgical procedure. To illustrate our implementation of these criteria, and the extremely severe degree of illness in patients operated on, consider the profile of the 32 patients who eventually committed suicide. All had major affective illness. 91% had 4 or greater major risk factors for suicide. The mean number of suicide risk factors was 6. The typical patient was someone who pre-cingulotomy had combined psychotic depression, anxiety, organic brain dysfunction, prior lethal suicide attempts, personality disorder, and family psychopathology. All of these patients were ambulatory, and were evaluated with their families by a neurosurgeon, neurologist, neuropsychologist and psychiatrist.

Personality change as a result of psychosurgery is the most commonly offered rationale for its effectiveness (Foltz 1962, White 1969, Ward 1948, Walsh 1977, Elithorn 1958, Koskoff 1948). Personality should be defined as the habitual patterns and qualities of behaviour of an individual as expressed by physical and mental activities and attitudes (Webster's Dictionary). Beliefs, attitudes and affective traits form this particular structure of mental traits we call personality. Affective traits are part of personality; affective states or moods are not. Mood, in response to stress or illness may vary discordant with personality. No where is this dysjunction more common or important than in affective illness where alien pathological mood states characterize illness, not true personality. This distinction is often not made explicit in discussions of affective change after psychosurgery. Improvement in mood does not mean a change in personality. The reduction of tortured self-concern, decreased rumination and worrying, impairment of the ability to form appropriate worried emotional responses, and a decreased social conscience have been offered as generalized personality changes that lie behind a mechanism of psychosurgery. Personality change is more than a simple theoretical question of mechanism because its status as the sine qua non of relief becomes critical in the overall assessment of benefits vs. risks. In other words if improvement is contingent on personality change, then this is a major adverse effect of psychosurgery. If on the other hand personality change is not intrinsic to the mechanism of psychosurgery, then this is quite reassuring. "Personality" profiles of anxiety and depression as described on MMPI scores are generally found to have been significantly lowered in successful cases. Koskoff 1948 and Freeman 1971 have stated that the beneficial effect of psychosurgery on pain remains only as long as the mental alterations can be demonstrated. They state that without these changes in personality, pain relief does not occur. This perspective is the predominant one, but comes from the earlier large frontal ablative procedures performed by Freeman *et al.* There is also some support for this point of view from the work of Foltz and White (1973) with more limited cingular lesions. However, there is evidence that personality change is not the sine qua non of efficacy. A change in the limbic suffering of affective disease or pain need not be associated with a change in personality traits, defined as both lifelong patterns of interaction with people and emotions. Hackett in 1969 analyzed 22 patients who had undergone staged medial leukotomy and found that severity of adverse personality change was not related to relief of pain (White 1969). Dieckmann (1987) reports on paedophiles who requested ventral-medial hypothalamotomy to undo what they saw as their undesirable sexual aggression. While this is a select motivated group it is also a disturbed group where sexual aggression has been repetitively severe and dangerously exploitative. In 8 patients they reported a decrease in sexual aggression, improved sexual relationship with a partner, and more openness, poise and egocentric feeling. Mindus and Myerson (1987) reported on effects of personality change with capsulotomy of which there are now 300 cases described in the literature. Rorschach results show no significant change in developmental level or the index of integration after capsulotomy. Affective parameters of hostility, somatic and psychic and muscular tension, and anxiety were decreased post-surgery. Their interpretation of these results is that the abeyance of anxiety results in a normalization of personality post-surgery. Walsh (1977) found that although self-concern, introversion, and depression decreased overall, particularly in the more improved cases, the items unchanged after surgery were the basic personality structure as reflected in the person's attitude to morality, sex, religion, and family. These reports suggest that true personality traits are not changed but rather state variables related to depressed affect. The most detailed neuropsychological study was that of Teuber and colleagues (1977) who after independent study of cingulotomy for pain and affective dysfunction commented on the lack of overall change in the person following surgery. They found it very difficult to find any tests that demonstrated a change in personality or cognition. In fact, intelligence quotients increased post-cingulotomy. The changes they found, primarily in older people, were in subtle cognitive skills, not in the overall personality. Patients had a decreased ability to draw complex figures, probably reflecting subtle impairment in conceptual thinking.

3. Altruistic vs. Exploitative Aspects of the Treatment

Is the purpose of the treatment to try and help someone or for some other secondary gain? Gross exploitation of patients must be exceptionally rare but has been recorded. Paul Lowinger describes that in 1972 while he was at a neuropsychiatric institute in Detroit he discovered a secret project for a series of experimental amygdalotomies on prisoner mental patients from a state mental hospital. The first subject was already being processed. He had been promised his freedom if he submitted to the procedure. The man chosen had committed a rape-murder and had been in a hospital for the criminally insane for 18 years during which time he had shown no aggressive behaviour.

Much more common are a number of psychological ethical traps or misadventures where there might be unconscious collusion between doctor and patient leading to inappropriate psychosurgery. The seriousness and desperation of the illnesses considered for psychosurgery make these problems particularly likely. Four examples of these problems are:

A. The physician may unconsciously identify with the anger or aggression of the patient, the doctor and patient embark on a folie a deux where the idea of "let's cut it out" has both concrete and metaphorical meanings.

B. Despondency over a very difficult situation might result in the doctor projecting his despair onto the patient. Doctor's own narcissism then becomes embroiled in proving that he at least is not helpless. This may result in unwise decisions for surgery where indications are not clear.

C. High frequency of attachment disorders in intractable depression and pain patients are strongly felt by both patient and doctor as a sense of recurrent loss. Intense attachments form in an effort to restore perceived loss, this attachment may be problematic if it is formed by way of a psychosurgical procedure. The restoration of attachment to some caring human relationship ought not to be done with the currency of surgery.

D. Denial of affect and personality problems is epidemic in many of the most severely ill patients. This may result in the patients selecting a few aspects of their problems and hiding many others that may be more germane. This easily leads to misdiagnosis of formal psychiatric disease.

4. Autonomy and Morality

The liberal obsession with not being morally judgmental, and being "for" human rights and autonomy is a common modern moral philosophy in the liberal tradition. Not being morally judgmental becomes a form of morality. However, this belief blurs the fact that effecting any moral judgement depends entirely on individual responsibility to each other. Moral judgement without responsibility is effectively amoral. This moral philosophy of being "for" autonomy and against the infringement of individual rights that a traditional conservative responsibility entails, is inherently powerless. In other words the promotion of autonomy, without consideration of the individual power plays required to implement and promote autonomy assures the continued failure of this amoral notion of human rights.

The libertarians' paradox is that they want "freedom" for others—as long as it is what the libertarian wants—and not if the person chooses another less liberal position. An example of this is when in the United States a convicted murderer wanted to be executed, but this wish was vetoed by the American Civil Liberties Union. Ethical questions about societal freedom to choose a particular course of action become questions about autonomy, morality, and responsibility. Societal "freedom" without responsibility becomes amoral and feckless at a societal level just as it does on an individual level.

5. The Nobility and Normality of Suffering

Suffering has always been part of human existence. The ability to alleviate suffering has created the ethical concerns about its value. In France in the 1700's physicians were burned at the stake for alleviating suffering with opiates, because it interfered with the spiritual importance of suffering. There are two extreme views. One is that suffering is noble, and intrinsic to the growth of humanity. The other view is that suffering is to be alleviated to almost any cost. The latter position is the typical teaching to the medical profession. The former opinion is held in a 1980 pastoral letter by Pope John Paul II regarding suffering. This stated that in Catholicism suffering was intrinsic to the maturational process, and therefore tolerable. However, this promotion of the nobility and normality of suffering implies that suffering is a cognitive matter. If suffering is a mentally ennobling experience, then suffering has become a matter of personal and interpersonal beneficence. This is not a new problem as seen in those that have advocated "beneficence" as the principal of morality in medicine. By turning suffering into an ennobling love problem, one is avoiding the difficult pragmatic questions about suffering, such as the despair and diminution of humanity that goes with it. This is analogous to the promotion of the moral principals of faith and justice as a reaction formation to unpalatable anti-intellectural and angry sentiments about authority. In fact the ethical problem of suffering is that difficult human emotions involved in suffering such as despair, anger, and love may be normal or unpalatable, or intolerable, or even lethal. These problems and contradictions are intrinsic to suffering, and cannot be resolved by edicts of any kind. Ethical consideration of these matters requires that at least the ethical questions be considered.

6. Politics

Political agendas are usually not far removed from ethical problems. Eliot Valenstein's 1986 book on the history of psychosurgery describes at length how important the political and personal processes were in the evolution of psychosurgery. In particular, the separation of psychiatry from neurologic medicine, administratiors from clinicians, and media predominating over scientific critique became major problems in the uncontrolled use of psychosurgery. Stanley Cobb, Harry Stack Sullivan, and Roy Grinker were but some of the medical scientists who criticized psychosurgery but were not heeded. Instead the politics of society as determined by the Reader's Digest, the Saturday Evening Post and human desperation dictated the course of events. Politics continue to actively intrude on the decision making process about psychosurgery. Two recent examples are the misrepresentation of Dr. John Donnelly in Kaplan and Sadocks' comprehensive textbook of psychiatry and T. Corwin Fleming in the Harvard Guide to Modern Psychiatry. Both authors have denounced the misquotation of their meaning by the U.S. Government Office of Health Technology Assessment (OHTA) and CHAMPUS respectively. Nevertheless these two organizations use these misquotations as information against psychosurgery. There is further misrepresentation of data in the health technology assessment reports on stereotactic cingulotomy as a means of psychosurgery, which was prepared by the U.S. Department of Health and Human Services under the auspices of Dr. Carter (1985). The studies cited in this report were overwhelmingly supportive of the efficacy of cingulotomy (11 out of 13 studies having been positive). Nevertheless, the conclusion of the report was negative, deemphasizing the positive studies and describing cingulotomy as an experimental procedure. At a January 1987 meeting with Dr. Carter and two of his associates were Nancy Cahill from the Duke Center for Health Policy Research and Education and Arthur Kobrine, neurosurgeon at George Washington University. Dr. Carter himself reread the report verbatim to reacquaint himself. He had no intention of considering reassessment, despite recognizing that he had misrepresented some information. He said that at best he might file an erratum. He would not answer or give any advice about an appeal based on a treatment of last resort. He said that efforts to enlist his advice were unheard. He reassured us that the report is not based on social questions. He said that speaking of economic issues was not appropriate. He contended that these were insufficient criteria for operating on these patients even while the report clearly lists them. Cingulotomy, according to Dr. Carter, is obscure enough and controversial enough to fail most appeals.

Acknowledgements

To Drs. Ballantine, Murray and Cassem for preliminary discussions about the subject of this manuscript, and C. Soreff for editorial assistance.

References

1. Ballantine HT, Bouckoms AJ, Thomas EK, Giriunas IE (1987) Treatment of psychiatric illness by stereotactic cingulotomy. Biological Psychiatry 22: 807–819
2. Bouckoms AJ (1984) Psychosurgery. In: Wall P (ed) Textbook of pain. Churchill Livingstone, Edinborough, pp 666–676
3. Dieckmann G (1987) The social outcome after ventromedial hypothalamotomy because of sexual violence. Paper presented at the Eurasian Academy of Neurological Surgery, Brussels. September 1987
4. Elithorn A, Glithero E, Salter E (1958) Leucotomy for pain. J Neurol Neurosurg Psychiatry 21: 249–260
5. Foltz EL, White Jr LE (1962) Pain "relief" by frontal cingulotomy. J Neurosurg 19: 89–100
6. Foltz EL, White LE (1973) Affective disorders involving pain. In: Youmans J (ed) Neurological surgery: a comprehensive reference guide to diagnosis and management of neurosurgical problems. WB Saunders, Philadelphia, p 1772
7. Freeman W (1971) Frontal lobotomy in early schizophrenia: long follow-up in 415 cases. Br J Psychiatry 119: 621–624
8. Kleinig J (1985) Ethical issues in psychosurgery. George Allen and Unwin, London
9. Koskoff YD, Dennis W, Lazovik D, Wheeler ET (1948) The psychological effects of frontal lobotomy performed for the alleviation of pain. Association for Research in Nervous and Mental Disease Proceedings 27: 723–753
10. Mindus P, Myerson B (1987) Aspects of personality in patients undergoing psychosurgical interventions. Paper presented at the Eurasian Academy of Neurological Surgery, Brussels. September, 1987
11. Office Health Technology Assessment: Stereotactic cingulotomy as a means of psychosurgery. 1985
12. Shevitz SA (1976) Psychosurgery: some current observations. Am J Psychiatry 133 (3): 266–270
13. Strom-Olsen R, Carlisle S (1971) Bi-frontal stereotactic tractotomy: a follow-up of its effects on 210 patients. Br J Psychiatry 118: 141–154
14. Sweet WH (1973) Treatment of medical intractable mental disease by limited frontal leucotomy—justifiable? NEJM 289: 1117–1125
15. Sweet WH (1980) Central mechanisms of chronic pain (neuralgias and certain other neurogenic pain). In: Bonica J (ed) Pain. Raven Press, New York, pp 287–303
16. Teuber HL, Corkin S, Twitchell TE (1977) A study of cingulotomy in man. Appendix to psychosurgery. Reported prepared for the National Commission for the Protection of Human Subjects of Biomedical and Behavioural Research. U.S. Dept. of Health, Education and Welfare Publ No (OS) 77-0002, 3: 1–115
17. Valenstein ES (1986) Great and desparate cures: the rise and decline of psychosurgery and other radical treatments for mental illness. Basic books, New York
18. Walsh KW (1977) Neuropsychological aspects of modified leucotomy. In: Sweet WH, Obrador S, Martin-Rodriguez JG (eds) Neurosurgical treatment in psychiatry, pain, and epilepsy. University Park Press, Baltimore, p 163
19. Ward Jr AA (1948) The anterior cingulate gyrus and personality. Association for Research in Nervous and Mental Disease Proceedings 27: 438–445
20. White JC, Sweet WH (1969) Pain and the neurosurgeon. Charles C Thomas, Springfield, Ill

Correspondence: A. J. Bouckoms, M.D., Harvard Medical School, Massachusetts General Hospital, Warren 6, Boston, MA 02114, U.S.A.

Acta Neurochirurgica, Suppl. 44, 179–180 (1988)

Ethics of Functional Neurosurgery

B. Ramamurthi*

Dr. A. Lakshmipathi Neurosurgical Centre, VHS Medical Center, Madras India

Summary

With the precision and accuracy available for modern techniques, operations on the brain for alleviating psychiatric disorders have become safer and reliable. Anxiety and feelings of doubt amongst the public about these procedures need to be dispelled. There is a need to compile accurate clinical and scientific data with careful follow up to assess the present status of functional neurosurgery and to ensure further progress.

Keywords: Psychosurgery; functional neurosurgery; ethics.

In the modern context and in view of the events and emotions observed in the USA and Japan against the practice of operating on the brain to relieve functional disorders, it has become necessary to state the case for functional neurosurgery, though in countries with a different background, this may appear to be stressing the obvious.

The efforts of scientists and medical men to improve the lot of their fellow beings have always initially met with opposition from a society which has not yet got used to a new idea. This has been so throughout the centuries, one of the later examples being the opposition to the introduction of anaesthesia for women during labour. Dogmas of ethics and religion are invoked to dampen the progress of science. This could happen in the twentieth century also despite the advent of the so-called permissive society. The objection to surgery on the brain to improve the lot of mentally afflicted patients, unfortunately, also belongs to this category of public resistance by the uninformed. Such a resistance which is expressed openly and sometimes violently in the developed countries has not had the beneficial result of adding caution to the surgeon's work, but has retarded or even stopped the progress of brain surgery in the field of mental disorders in these countries. The ingrained ancient concepts of the mind being separate from the body and of its inviolability and sanctity still seem to be a drag on an unbiased approach to the problem of the afflicted.

The medical profession has taken upon itself the task of alleviating many ills and afflictions of the body. Similarly it is its duty by all possible means at its disposal, to help the mentally afflicted since mental diseases are also basically due to an underlying brain disorder, chemical or organic. Consequently, success in restoring the mental balance of a person and returning him to society is as much the aim of surgeons and physicians as their declared and accepted ambition to treat bodily ailments such as hypertension, diabetes or cancer. The dichotomy of approach to a mental problem as opposed to a physical problem clouds judgement and thus prevents desirable relief being given to mentally afflicted persons who are in crying need of help to restore them to the normal stream of life, by whatever means available. To deny them this is to deny the role of relief and succour to medicine.

It is well-known, that operations on the brain, if they misfire, may after the personality of the patient, but what seems to be missed and not mentioned is the fact that precise neurosurgical procedures in appropriate cases enable the patient to re-establish his place in society without any alteration in his basic personality. Moreover a person who becomes aggressive or violent against his desire or who cannot get out of his depression in spite of his efforts has a right to demand a return to normal personality. In such cases alteration in the diseased personality is the aim of treatment, this alteration being towards the better, to help the patient regain his place in society. The argument repeated endlessly that personality should not be altered seems to

* Published in the Text Book of Neurosurgery. Ramamurthi B, Tandon P (eds) (1980) Orient Longman Ltd, New Delhi, pp 1120–1122

be based on a concept that personality is something sacrosanct and should not be changed. This reminds one of the mediaeval periods when it was argued that poverty was bestowed by divine will and should not be altered. The relics of fear once caused by the original prefrontal leucotomy continue to haunt the thinking on the problem and even the very word psychosurgery seems to rouse deep-seated dread of creating morons and robots from human beings. On this topic science fiction has done more harm than good to the cause of the mentally afflicted. The term psychosurgery is indeed unfortunate as one does not operate on the psyche and a much better term is functional neurosurgery.

The fact that the operations are on the brain, "the seat of the mind" and the confusion between the soul and the mind in some aspects of Western thought seem to have raised a furore over psychosurgical procedures. Luckily, in the Orient, such a mix-up is not seen, the mind being associated only with the physical body at a much lower level than the soul. Disturbances, chemical or otherwise in the neuronal circuits cause disturbance in mental function. With the modern tools available in neurosurgery, it is possible to disconnect malfunctioning circuits enabling the brain to re-establish normality. If the resultant alteration in brain function is towards betterment and towards enabling the patient ro rejoin society, then functional neurosurgery has a case. Reports from a few centres in the world, where such procedures are adopted and the patients are followed up with care show that with precise and well planned procedures in properly selected patients, the results have been extremely satisfactory with no deleterious effect on personality. Minor alterations picked up by extensive and often unnatural psychological testing should not be allowed to cloud the fact of resultant benefit to the patient, his family and society.

The public and psychiatrists in India are desirous of taking advantage of precise neurosurgical procedures in well selected cases. The results have encouraged the surgeons to pose to the psychiatrists the question of earlier referral. So far the cases operated upon have been those who had unsuccessfully undergone psychiatric therapy for many years. It is worth while investigating what may be the optimum time for referral for operation, to avoid unnecessary and prolonged travail for the patient and the family and also to ease the burden on the psychiatrists and the hospitals.

To quote Bullock (1975): "A dispassionate examination of the possibilities of precise surgical procedures on the brain is called for, especially for the relief of specific psychiatric conditions like schizophrenia, obsession, addictions, aggression and violence. From the perspective I have gained, as a non-physician neuroscientist, the most widespread criticism of psychosurgery is not the abuse in a limited number of cases but the serious shortcoming in a large number of cases in respect to pre-operative and post-operative evaluation. It is clear that the climate of today in this part of the world demands a marked increase in the detail, the time-span, and the multidisciplinary character of evaluation, with the corollary requirements for the experts beyond those minimally needed for surgical and medical care—in particular for objective, professional assessment of social behaviour. Every patient should be properly evaluated, followed-up and reported. Fortunately, this is the fact of the problem most amenable to immediate solution. The bias against functional neurosurgery based upon the untoward results of the original leucotomy operation, the confusion of the idea of the soul with the mind and unjustified fears about the personality changes—all these need to be carefully dispelled by a calm analysis of the indications and results of psychosurgery as is being done in the other realms of surgery and their publication for informed criticism. Only this can make functional neurosurgery available to those who need it".

Reference

1. Bullock TH (1975) Neuroscientists on psychosurgery. Arch Neurol 32: 73

Correspondence: Prof. B. Ramamurthi, M.D., Head of the Department of Neurosurgery, D.A. Lakshmipathi Neurosurgical Center, VHS Medical Center, Madras 113, India.

VIII. General Conclusions

Acta Neurochirurgica, Suppl. 44, 183–185 (1988)

General Conclusions

J. Brihaye

Bruxelles, Belgium

To draw conclusions from this meeting is really a perilous task owing to the fact that neurosurgeons in general are not familiar with several matters which were here considered. Therefore, I will limit myself to some comments which occurred to me while listening to the speakers.

The lectures delivered by the philosophers on the concept of personality were indeed very stimulating and made obvious that, in addition to biophysical data, there was a noticeable impact from the environment in making up each individual personality. In fact, the multifactorial make-up of the person is evident but the relative importance of the various factors at work is differently appreciated by everyone according to his philosophical or religious convictions. It therefore goes without saying that the concepts and theories developed by the philosophers, as well as the scientific lectures, are not to be mistaken as official statements of the Academia itself and the Academy in any way.

From the philosopher's analysis, as well as from the essay of Huber on the assessment of personality, it clearly appears that the concept of person is not a static and congealed state; on the contrary they all point out that a large and diverse approach was necessary and that interaction of factors and reaction to various punctual situations were operative in the dynamics of personality. The difficulty to assess, in neurological diseases and in neurological surgery, a change of personality instead of a simple disturbance of behaviour deserved to be stressed with regard to the appraisal of results following neurosurgical procedures.

Particularly in neurotraumatology many clinical studies have shown that severe brain lesions may definitely alter the personality of the injured; the changes are more or less important and subject to evolution during several months, even during several years. The case presentation by van Dongen and Arts of an adolescent who partially recovered after two and half years of vegetative state clearly indicate the potentiality for recovery of brains still immature. This case demonstrates how carefully we have to be when deciding upon the outcome of the injured patients. Brooks in his lecture is right in emphasizing the real difficulty which still persists about the prediction of traumatic consequences on the personality. Thiery also lay stress on this point, but he also reminded us that we must also take into account the quality of life of the patient. Our duty consists not only in saving the life but at the same time in restoring the best physical and mental possibilities, in order that the patient could appreciate the life itself. In the case reported by van Dongen and Arts, the mental recovery, in spite of the improvement obtained, remains at a lower level. The relation of this exemplary case also compels us to consider carefully the social and economic burden of taking on the care of such cases. Brooks, with great clarity, discussed the family concern at these cases, relating the devastating effect on the family life that heavy physical and mental handicaps may provoke. Of course there is no rule in such matters and each injured person has to be considered in all his medical and psycho-social characteristics. However, there is no doubt that, also in this field of severe traumatic lesion of the brain, the personal philosophical background of the physician and his clinical experience will put pressure on his decision making.

Seron and Van der Linden with reason criticize the numerous clinical studies which appeared in the literature; they also call attention to the great difficulty in these studies of appreciating the disorders of the personality because of the multifarious characteristics at work and the lack of scientific objectivation and correlation. They have observed, as did Yvan Lebrun, that the impact on the behaviour of disturbances in communication and verbal performances may often be interpreted as resulting from psychological disorders more than from a localized brain lesion. Our clinical experience gained during a long association with Yvan Lebrun, in his department of neuro-linguistics, allows us also to emphasize that, in addition to primary or secondary psychological disturbances, the difficulty or the loss of communication with his environment experienced by the aphasic patient has a direct influence upon his behaviour and very probably upon his personality. The language indeed is not only a simple mean to communicate, but it also represents a direct objectivation of the personality.

Trimble on the one hand and Gillingham on the other, have a large experience relating to epilepsy and personality. Both of them have observed undeniable changes of personality correlated with epilepsy, mainly with temporal lobe pathology. At the same time, both of them have noticed a striking improvement of personality disturbances by successfully treating the epilepsy, the improvement being in direct relation to the disappearance or reduction of the epileptic fits.

Several lectures in keeping with the surgical treatment of mental and behaviour disorders indicate that there undoubtedly still exists a large consternation over its correct indication. The multiplicity of the targets used by the psychiatric surgeons would make one think that there remains insufficient rationale in the clinical decision. M. Nys told us that even the definition of psychosurgery is not yet clearly worked out and in addition he notices that psychosurgery is mainly performed in the Nordic countries and hardly ever done in Latin-oriented countries. Such a difference between Nordic, mainly protestant countries and South, mainly catholic ones has also been stressed by historians and economic experts in other fields of human activity. Ramamurthi, in the notes he circulated among the participants, pointed out another discrepancy between the Oriental countries where psychosurgery is currently performed and Western countries where it is irregularly done: "That operations on the brain, if they misfire, may alter the personality of the patient is well-known, but what seems to be missed and not mentioned is the fact that precise neurosurgical procedures in appropriate cases enable the patient to re-establish his place in society without any alteration in his basic personality. Moreover a person who becomes aggressive or violent against his desire or who cannot get out of his depression in spite of his efforts has a right to demand a return to normal personality. In such cases alteration in the diseased personality is the aim of treatment, this alteration being towards the better, to help the patients regain his place in society. The argument repeated endlessly that personality should not be altered seems to be based on a concept that personality is something sacrosanct and should not be changed ..." The fact that the operations are on the brain, "the seat of the mind" and the confusion between the soul and the mind in some aspects of Western thought seem to have raised a furore over psychosurgical procedures. Luckily, in the Orient, such a mix-up is not seen the mind being associated only with the physical body at a much lower level than the soul."

These remarks made by Ramamurthi have to be brought nearer to the statement of Bouckoms about psychosurgery considered as a problem of moral philosophy. Bouckoms indeed, while examining the efficacy and the morality of psychosurgery, considers that psychosurgery is only acceptable if it does not provoke a change of personality.

It appeared evident from these various contributions that guidelines for psychosurgery ought to be drawn up, in spite of the fact that excellent results with restoration of a normal life have in some cases only been obtained by surgery on the brain. The fear of a potential misuse of psychosurgery as a agent for social or political control still exists. In addition, the reluctance to definitively and deliberately destroy nervous structures within the brain can be understood. In my opinion, and it seems to me that there was a general agreement on that, the fundamental rule for the doctor is to produce the best care for the patient; when surgery on the brain is regarded as the best treatment in some specific mental or behaviour disorders, then the surgical procedure has to be performed.

This basic rule being accepted, there still remain sensitive problems to solve; I will simply quote a few of them:

— is it necessary to obtain an informed consent of the patient for psychosurgery? Nys and Bouckoms have with reason insisted upon this problem of autonomy. It stands to reason that it is better to have this patient's consent. But it cannot be an absolute requirement because, if it is the case, a number of institutionalized persons thereby would be denied treatment that would be beneficial. In fact, whatever the priority given today to the principle of autonomy, the neurosurgeon has the duty to remain a decider for the benefit of his patient.

— Is psychosurgery for children acceptable? The answer ought to be positive when psychosurgery appears to be the only solution to help for example an agitated and hyperactive child to be better integrated and accepted in the family circle.

— Psychosurgery for sex offenders in order to reduce their sexual desire remains debatable; surgical castration is regarded by some as more advisable than hypothalamotomy because they consider testicles less "noble" than the brain. Taking up the words of Per Mindus, I am asking if there is not a psychological diplopia with regard to these two surgical procedures. Dieckman besides demonstrated valuable results with this specific psychosurgical procedure and I don't see a real basic difference between castration and hypothalamotomy. However in these cases the indication for surgical treatment has to be considered after full expertise and with the full consent of the informed patient.

— Is it a requirement for the neurosurgeon to make use of consulting colleagues, psychologist and/or psychiatrist, before deciding to operate on the brain? Personally I think that he has to fulfull this requirement everytime not only by fear of lawsuits or erroneous allegations but mainly in order to offer to the patient all the guarantee of the best therapeutic measure.

Finally, we have to consider psychosurgery as a functional neurosurgery in the frame of which its place is to be found at the side of surgery for pain or involuntary movements; actually its impact on personality is generally denied and I will quote Sweet who declares that psychosurgery reduces undesirable behaviour but does not alter the patient's fundamental personality.

Correspondence: J. Brihaye, M.D., Prof. Emeritus of Neurosurgery, Avenue des Franciscains, 98, B-1150 Bruxelles, Belgium.